STATISTICAL ANALYSIS
OF MEASUREMENT ERRORS

EXXON MONOGRAPHS

STATISTICAL ANALYSIS
OF MEASUREMENT ERRORS

JOHN L. JAECH

An Exxon Monograph

JOHN WILEY & SONS
New York · Chichester · Brisbane · Toronto · Singapore

Copyright © 1985
Published by John Wiley & Sons, Inc.

All rights reserved. Published simultaneously in Canada.

Reproduction or translation of any part of this work
beyond that permitted by Section 107 or 108 of the
1976 United States Copyright Act without the permission
of the copyright owner is unlawful. Requests for
permission or further information should be addressed to
the Permissions Department, John Wiley & Sons, Inc.

This publication is designed to provide accurate and
authoritative information in regard to the subject
matter covered. It is sold with the understanding that
the publisher is not engaged in rendering legal, accounting,
or other professional service. If legal advice or other
expert assistance is required, the services of a competent
professional person should be sought. *From a Declaration
of Principles jointly adopted by a Committee of the
American Bar Association and a Committee of Publishers.*

Library of Congress Cataloging in Publication Data:

Jaech, John L.
    Statistical analysis of measurement errors.

    "An Exxon monograph."
    Includes index.
    1. Errors, Theory of.   2. Mensuration.   3. Estimation
theory.   I. Title.

QA275.J34 1985   511'.43   85-5357
ISBN 0-471-82731-2

Printed in the United States of America

10 9 8 7 6 5 4 3 2 1

To Alyce

FOREWORD

The nature of Exxon Nuclear's business of fabricating uranium fuels for commercial power reactors requires replicate measurements of difficult-to-measure characteristics. Errors of measurement are of dominant importance in the Company's quality assurance program and in implementing special nuclear materials safeguards. This Exxon Monograph provides a comprehensive, user-oriented approach to the statistical analysis of replicate measurement data.

In this book the author, John L. Jaech, has developed a general model of statistical inference based on maximum likelihood for experimental situations in which a multiple number of items are each measured by several different measurement methods. While the method of maximum likelihood has been well known to have optimum statistical properties, its application to multiple item, multiple measurement situations has been limited by the unavailability of computation methods. The method of maximum likelihood is compared to other methods of estimation with the advantages and disadvantages of each pointed out. Experience with the use of the maximum likelihood method is demonstrated with numerous examples.

Although the method of maximum likelihood in this problem context was developed for use in the nuclear field, the subject matter of this monograph is by no means limited to this field of application. Rather, it has broad application in manufacturing and research facilities, wherever measurements are performed, and is certainly of importance to the petroleum and chemical industries. We hope that by sharing the methods developed in this monograph, others will be able to further improve the precision and accuracy of their measurement methods.

R. W. McCullugh
Vice-President
Engineering and Production
Exxon Nuclear Company, Inc.

PREFACE

For the past fifteen years of my career as a statistical consultant in the nuclear industry, I have been heavily involved in an area in which errors of measurement are crucial: the safeguarding of special nuclear materials. "Safeguards" is a collective term comprising measures used to guard against the diversion of special nuclear materials from uses permitted by law or treaty.

An important safeguard involves maintaining records of receipts, shipments, waste streams, and periodic physical inventories and verifying these records through inspection. This process requires the measurement of difficult-to-measure materials. In reaching decisions about the acceptability of a given material balance, errors of measurement play a major role; thus it is important to characterize such errors properly—to obtain valid estimates of the many error parameters and, more generally, to apply good statistical techniques when making inferences about them.

In considering these problems of statistical inference, there are several variants of what may be considered a general experimental situation. "Paired data" are quite common; this term is used to describe the measurement of the same item by two measurement methods. "Methods" may, on the one hand, refer to the same analytical technique, possibly involving different analysts or pieces of equipment, or even different laboratories, or it may refer to totally different measurement techniques. Paired data occur when items are transferred between two facilities, with each facility performing measurements; when an internal audit of a facility is conducted to verify previously measured values; when a third-party performs the verification inspection, and when characterizing or evaluating different measurement techniques.

In safeguards practices, it is also common for the same items to be measured by three or more measurement methods. Items are often transferred among facilities to compare measurements. Even within a facility, it is common practice to perform replicate measurements by different techniques. As new measurement techniques are developed, comparisons are made for evaluative purposes with existing techniques.

All of these practices can be summarized in the general experimental situation in which n items are each measured for the same characteristic by each of N measurement methods. The data form an $n \times N$ matrix. For some of the methods, replicate measurements may be performed, but only the mean for each item is reported and used in the statistical analysis. A distinction is made between a random and a fixed model. In a random model, items selected for measurement are drawn randomly from a population in which the random variable is normally distributed; in a fixed model the items are preselected to span some given range.

In considering this experimental situation, I have found various inferential statistical techniques scattered throughout the literature. Some are commonly known and applied; others are not. In practice, I have found that competing methods of point estimation, for example, may give quite different results. In one such example the likelihood corresponding to the point estimates produced by an estimation method in common usage was disturbingly smaller than that corresponding to a null hypothesis of equal parameter values. It was this example that prompted me to focus on maximum likelihood estimation as a preferred approach. Thus the common thread running through the variants of the general model discussed in this book is statistical inference based on maximum likelihood.

However, the focus in this book on maximum likelihood is not to the exclusion of other methods of estimation. I feel that it is important to include a discussion of alternative statistical techniques, especially when alternative techniques are in frequent use because of their relative simplicity. On occasion, efficiency may well be compromised for simplicity. Thus I have presented the various methods of estimation, pointing out the advantages and disadvantages, but leave it to the user to choose the method for a given case.

After an introductory chapter included to make this book somewhat self-contained, succeeding chapters develop the general experimental situation with increasing complexity. In Chapter 2, I consider the case in which the measurement methods have equal precisions, while in later chapters, the models assume unequal precisions. Chapter 3 deals with $N = 2$ measurement methods. In Chapters 4 and 5, I deal with inference based on paired differences, for $N = 3$ in Chapter 4 and for $N \geq 4$ in Chapter 5. The original data form the basis for the statistical inference in Chapter 6, for $N \geq 3$. A comparison of moment and maximum likelihood estimators is found in Chapter 7, while Chapter 8 gives a bibliography of works not referenced in earlier chapters. In Chapter 9 I give an example of outputs from the computer programs, and also give program listings.

Although the choice of material in this book was motivated primarily by my work in the specialized area of nuclear materials safeguards, the statistical techniques have much broader application. I have written primarily for the engineer and physical scientist because this has been the emphasis in my career. All the examples are from this field. However, the material may be of interest to social scientists as well. Specifically, the model in Chapter 6 is a common model in psychology and sociology for which a flexible and general computer program, LISREL VI of the Software Package for the Social Sciences (SPSS), may be used.

In the developing years of statistical inference, the application of maximum likelihood estimation methods, both in the general situation covered in this book and elsewhere, was hampered by computational complexity. For ease in application, simpler statistical techniques were developed first; as such, they continue to be widely used. Now, with the broad availability of high speed computers, and, indeed, with the ever-increasing capabilities of desk-top computers and even of hand-held calculators, this hindrance has become less important. To facilitate use of maximum likelihood estimation, computer programs for use with the estimation techniques that are presented in this book are included. However, it would not be difficult for potential users to develop their programs, possibly for use on desk-top computers. In fact, it is not out of the question for users with smaller dimensioned problems to obtain estimates of the parameters using hand calculators.

Many of the techniques described in this book are scattered in one form or another throughout the literature. References cited are listed at the end of each chapter and, as stated earlier, a more complete bibliography is included as Chapter 8. (The reader will note that most of the cited works are from the literature familiar to those working in the engineering and physical science fields, for reasons given earlier.) Because of the scattered references, I feel that this book fills a need—presenting a complete exposition of an important problem.

I emphasize that the subject of statistical analyses of measurement errors is a broad one, and in order to keep the book of manageable size, I had to omit related topics of interest. I have not included any results wherein inferences on measurement error parameters are based on direct analyses of the replicate data. Also, I have omitted any mention of Bayesian estimation. Neither omission is intended to minimize the importance of methods of inference based on such approaches.

My purpose has been to make available to engineers and physical scientists a number of methods that may be used for statistical inference in the experimental situation in which n items are measured for the same characteristics by N measurement methods. The book is intended for self-study and as a reference book. No previous formal training in statistics is required, but some familiarity with statistical concepts is certainly helpful. To aid the reader interested primarily in applications, I defer most proofs to appendices, which are included for those readers who choose to delve further.

JOHN L. JAECH
Consultant, Statistics/Safeguards
Exxon Nuclear Company, Inc.

Bellevue, Washington
June 1985

ACKNOWLEDGMENTS

Although I accept sole responsibility for any imperfections in this book, I cannot claim sole credit for its production. Extraordinary dedication to the typing of the several versions of the manuscript, replete with tables and mathematical expressions, was exhibited by Marilyn Strickwerda, to whom I am most grateful. I also owe a debt of gratitude to Joyce Barrett, who performed the editing task, in many instances conveying the message that I had intended, but in a simpler and more direct manner. To the unknown reviewers I am grateful for their suggestions, which resulted in major improvements between the initial and final drafts. The computer programs were written by Sharon Peterson, whose contribution is gratefully acknowledged. I very much appreciate the opportunity to write this book provided me by Exxon Nuclear Company, first in proposing the project and then in making available to me the resources needed to complete it. In particular, George Sofer and Sam Beard provided the initial support and encouragement, and Bob McCullugh continued this support as the work progressed. Finally, I acknowledge the continued encouragement given me by my wife, Alyce, who is as great a fan of mine as I am of her.

J.L.J.

ABOUT THE AUTHOR

John L. Jaech joined Exxon Nuclear Company in 1970 following previous employment with the General Electric Company and with Battelle Laboratories. He has been working in statistical applications in the nuclear field since 1953 following receipt of his Master of Science Degree in Mathematical Statistics from the University of Washington. Mr. Jaech has served as Consultant to the International Atomic Energy Agency in Vienna on a number of occasions dating back to 1972. He is the author of the book, *Statistical Methods in Nuclear Materials Control*, has authored individual chapters in several other books, has published almost 50 papers on applied statistics in a number of technical journals, and has delivered an equal number of papers at professional society meetings and symposia. Mr. Jaech is Past Chairman of the Institute of Nuclear Materials Management, an international body of professionals engaged in nuclear materials safeguards.

CONTENTS

STATISTICAL ANALYSIS
OF MEASUREMENT ERRORS

Chapter 1

INTRODUCTION

1.1 PURPOSE

Beginning with Chapter 2, this book focuses on a rather sharply defined experimental situation.

> *Experimental Situation.* A number of items (n) are each measured once for the same characteristic by each of N measurement methods. The value of this characteristic for a given item does not change during the experiment. Thus, the data consist of nN observations.

In this description, reference is made to N measurement "methods." This term is purposely vague: "Methods" may refer to quite different ways of measuring a given characteristic, such as measuring the density of a nuclear reactor fuel pellet by geometric means or by an immersion method. The term may also refer to the same method of measurement carried out by different instruments or pieces of equipment or possibly different technicians. When chemical analyses are performed in different laboratories, each laboratory is identified as a different measurement "method," as the term is used here. In a given application, the definition of the method will influence the construction of the mathematical model and the accompanying assumptions.

The statistical problem to be covered in the remaining chapters of this book is that of estimating the parameters of the model, obtaining both point and interval estimates. Problems associated with testing hypotheses about the parameters are also considered.

Before discussion of this principal topic, the remaining sections of Chapter 1 present basic introductory material. This presentation makes the book largely self-contained so that a reader with little prior knowledge of statistics will be able to understand the later chapters. For those readers with some background in statistics, Chapter 1 provides a concise review of concepts and results used in the remainder of the book. While some parts of Chapter 1 may not be needed specifically for studying the succeeding chapters, they are included in order to provide a logical development of results. A reader well versed in statistics may want to begin with Chapter 2.

1.2 DESCRIPTIVE STATISTICS AND INFERENTIAL STATISTICS

It is important, at the start, to distinguish two classes of statistical methods. Some statistical methods are used to describe or summarize the information content in a set of data with no attempt to generalize this information beyond the data. This class of methods is called *descriptive* statistics. The methods of descriptive statistics are well known and commonly used. They include many types of statistical graphs, charts, tables, and indices. Examples would include batting averages in baseball, the Dow-Jones stock market average in economics, coastal area tide tables, and pie charts of expenditures, to name a few of the many common ones.

This book is concerned with the second class of methods, labeled *inferential* statistics. In inferential statistics, one makes inferences about a larger amount of possible data based on an existing data set, presumed to be representative in some sense of the larger set.

The methods of inferential statistics are not as familiar as those of descriptive statistics. They are based on a thought process that may be rather foreign to some. Commonly, one may be able to apply selected methods of inferential statistics, yet not have a full understanding of their rationale. Such an understanding is best gained by experience, and it is this lack of understanding that may lead to misapplications of statistical methods. Proper application of inferential statistics leaves little room for ambiguity in thinking. For this reason, considerable attention is given in this book to the mathematical model that forms the basis for the statistical methods presented here.

As a final note, some of the techniques used in descriptive statistics are identical, to a point, with those used in statistical inference. For example, in both applications, one might calculate the average of some set of numbers or create tables or plot the data. With descriptive statistics, such exercises lead directly to the desired result—some kind of data summary. In inferential statistics, these same techniques lead to intermediate results, which then form the basis for statistical statements about a larger population of data.

1.3 PROBABILITY

Methods of statistical inference are grounded in the theory of probability. Mathematicians develop probability concepts axiomatically, but for our purposes it is more instructive to develop these concepts on a more intuitive basis.

1.3.1 Probability Concepts

Probability is a measure of the likelihood that an event will occur. This measure may be defined based on three interpretations: the classical interpretation, the frequency interpretation, and the subjective interpretation.

Classical Interpretation of Probability. The laws of probability (Section 1.3.2) are most easily understood if one has in mind the classical interpretation of probability. However, the laws apply with equal validity to all three interpretations of probability considered here.

The classical interpretation of probability says that if an event or experimental outcome can occur in N equally likely and different ways, and if n of these have some defined attribute A, then

$$\Pr(A) = \frac{n}{N} \qquad (1.3.1)$$

where $\Pr(A)$ denotes the probability of the occurrence of event A.

Note that since n may range from 0 to N, $\Pr(A)$ may range from 0 to 1. If $\Pr(A) = 0$, the event A cannot occur, while if $\Pr(A) = 1$, it is certain to occur. If A and B denote two events, then $\Pr(A) < \Pr(B)$ means that the event A is less likely to occur than the event B.

Calculations of probabilities in games of chance are often based on this classical interpretation of probability. It is noted, however, that although Eq. 1.3.1 appears to be a simple formula, applying it to calculations of probabilities may be quite complex because of difficulties in assigning values to n or N. A noted American mathematician and logician, Charles Sanders Peirce, once observed that in no other branch of mathematics is it so easy for experts to blunder as in probability theory, and the incidences of such blunders throughout history support this observation. (This note is included as a source of comfort to those readers who wrestle unsuccessfully with seemingly innocent problems in probability.)

EXAMPLE 1.A

(a) In throwing four dice, what is the probability that no aces (ones) appear? (b) In throwing two dice 24 times, what is the probability that no pairs of aces appear?

(a) For a single die, $N = 6$ and $n = 5$. For the four dice, $N = 6^4$ and $n = 5^4$. Therefore, if A is the event that no aces appear,

$$\Pr(A) = \frac{5^4}{6^4} = 0.4823$$

(b) For a single throw of the pair of dice, $N = 36$ and $n = 35$. For the 24 throws, $N = 36^{24}$ and $n = 35^{24}$. If B is the event that no pairs of ones appear, then

$$Pr(B) = \frac{35^{24}}{36^{24}} = 0.5086$$

Historical Note.   Chevalier de Méré, a gambler of some note, thought that the two probabilities ought to be equal and blamed mathematics for his gambling losses (1).   ∎

EXAMPLE 1.B

(a) A container of spare parts in inventory contains 10 items, 2 of which are defective. One spare part is selected at random. What is the probability it is defective?

Here, with A being the event that the item drawn is defective, $N = 10$, $n = 2$, and $Pr(A) = 0.2$.

(b) Three parts are selected at random. What is the probability that both defectives are included in this sample of three parts? Keep a record of the order in which the three parts are drawn. Then, $N = (10)(9)(8) = 720$. Labeling x as a defective part and o as a good part, the sequences of drawings that result in two defects are

$$xxo, xox, oxx$$

The number of sequences: xxo is $(2)(1)(8) = 16$. The number of sequences: xox is $(2)(8)(1) = 16$. The number of sequences: oxx is $(8)(2)(1) = 16$. Therefore,

$$n = (3)(16) = 48$$

$$Pr(A) = \frac{48}{720} = \frac{1}{15}$$ ∎

Frequency Interpretation of Probability.   Consider a situation in which an experiment is conducted N times, and a defined event A occurs n times. Then the limit of n/N as N gets large is defined to be $Pr(A)$, the probability that the event A will occur.

EXAMPLE 1.C

A standard sample containing a nominal 29.27% iron is assayed by a number of students in the chemistry laboratory of a large university. Letting A be the event that the assayed value is less than 29.40%, the following table gives values for $\Pr(A)$ with increasing N. The results are tabulated after each set of five results is reported to the instructor.

N	n	$\Pr(A)$
5	3	0.60
10	8	0.80
15	10	0.67
20	14	0.70
25	18	0.72
30	23	0.77
35	27	0.77
40	30	0.75

∎

Subjective Interpretation of Probability. Management decisions are based, at least partly, on a manager's degree of belief about the likelihood that a specified event A will occur. This may be thought of as a quantification of professional judgment based on experience in that it expresses the betting odds for a given proposition. Odds of 4 to 1, say, correspond to a probability of 0.8. In general,

$$\text{odds} = \frac{\text{probability}}{1 - \text{probability}}$$

1.3.2 Laws of Probability

Although the laws of probability are most easily understood under the classical interpretation of probability, they are equally valid under all interpretations. These laws are presented here as rules. (Note that other notation, such as set theory, may be used.)

Rule 1. If A and $\bar{A}$ denote, respectively, the occurrence and the nonoccurrence of an event, then

$$\Pr(A) + \Pr(\bar{A}) = 1 \tag{1.3.2}$$

or equivalently,

$$\Pr(\bar{A}) = 1 - \Pr(A) \tag{1.3.3}$$

To see the truth of this rule, from Eq. 1.3.1, since n of the N outcomes have the attribute A, then $(N - n)$ have the attribute $\bar{A}$ and hence

$$\Pr(\bar{A}) = \frac{N - n}{N} = 1 - \frac{n}{N} = 1 - \Pr(A)$$

Rule 2. If A and B are two independent events in that the occurrence or nonoccurrence of A in no way affects the occurrence or nonoccurrence of B, and if AB denotes the occurrence of both events, then

$$\Pr(AB) = \Pr(A)\Pr(B) \qquad (1.3.4)$$

EXAMPLE 1.D

In Example 1.C, find the probability that two successive students obtain values less than 29.40% iron. Use $\Pr(A) = \Pr(B) = 0.75$. Then from Eq. 1.3.4,

$$\Pr(AB) = (0.75)(0.75) = 0.5625$$

To extend Eq. 1.3.4, the probability that five successive students will obtain values less than 29.40% iron is

$$\Pr(ABCDE) = (0.75)^5 = 0.2373$$

One assumes that the students obtain their values independently; that is, no student is guided in any way by the results of his or her classmates. ∎

Rule 3. The probability that either A or B or both will occur is written $\Pr(A + B)$, and is given by

$$\Pr(A + B) = \Pr(A) + \Pr(B) - \Pr(AB) \qquad (1.3.5)$$

If $\Pr(AB) = 0$, then A and B are said to be *mutually exclusive* events.

EXAMPLE 1.E

In continuing the example under Rule 2, the probability that either the first or the second or both students obtain values less than 29.40% iron is, by Eq. 1.3.5,

$$\Pr(A + B) = 0.75 + 0.75 - 0.5625 = 0.9375 \qquad ∎$$

Rule 4. If A and B are not independent events, then

$$\Pr(AB) = \Pr(A)\Pr(B|A)$$

$$= \Pr(B)\Pr(A|B) \qquad (1.3.6)$$

where $\Pr(B|A)$ is the conditional probability that B will occur, given that A has occurred, and $\Pr(A|B)$ is similarly defined. Note that when A and B are independent events, then $\Pr(B|A) = \Pr(B)$, and Rule 4 reduces to Rule 2.

EXAMPLE 1.F

Suppose that 100 items are weighed, 40 on scale 1 and 60 on scale 2. On scale 1, the weighing error exceeds 5 grams 25% of the time, while on scale 2 errors of this magnitude occur 50% of the time. Define the events:

A: item is weighed on scale 1

$\overline{A}$:   item is weighed on scale 2

B: weighing error exceeds 5 grams

The given probabilities are

$$\Pr(A) = 0.4$$

$$\Pr(\overline{A}) = 0.6$$

$$\Pr(B|A) = 0.25$$

$$\Pr(B|\overline{A}) = 0.50$$

Find the probability that an item selected at random is weighed on scale 1 and has a recorded weight in error by more than 5 grams. This is the event AB. From Eq. 1.3.6,

$$\Pr(AB) = (0.4)(0.25) = 0.10$$

Find the probability that an item selected at random has a weighing error in excess of 5 grams. This is the event B. From Eq. 1.3.5, noting that the events AB and $\overline{A}B$ are mutually exclusive,

$$\Pr(B) = \Pr(AB + \overline{A}B) = \Pr(AB) + \Pr(\overline{A}B)$$

$$= 0.10 + \Pr(\overline{A}B)$$

$$\Pr(\overline{A}B) = \Pr(\overline{A})\Pr(B|\overline{A}) = (0.60)(0.50) = 0.30$$

Therefore,

$$\Pr(B) = 0.10 + 0.30 = 0.40$$

Given that the weighing error for a given item exceeds 5 grams, find the probability that the item was weighed on scale 1. This is the event $A|B$. From

Eq. 1.3.6,

$$Pr(A|B) = \frac{Pr(AB)}{Pr(B)}$$

$$= \frac{0.10}{0.40} = 0.25$$

(See also Example 1.G.)   ∎

The four rules given here can be extended to include more than just one or two events. The multievent equivalents of the four rules are as follows.

Rule 1a.   If the events $A_1, A_2, \ldots, A_n$ are mutually exclusive, but if one of the events must occur, then

$$\sum_{i=1}^{n} Pr(A_i) = 1 \tag{1.3.7}$$

Rule 2a.   For independent events $A_1, A_2, \ldots, A_n$,

$$Pr(A_1 A_2 \ldots A_n) = \prod_{i=1}^{n} Pr(A_i) \tag{1.3.8}$$

Rule 3a.   For three events A_1, A_2, and A_3,

$$Pr(A_1 + A_2 + A_3) = Pr(A_1) + Pr(A_2) + Pr(A_3)$$

$$- Pr(A_1 A_2) - Pr(A_1 A_3)$$

$$- Pr(A_2 A_3) + Pr(A_1 A_2 A_3) \tag{1.3.9}$$

For n events $A_1, A_2, \ldots, A_n$,

$$Pr(A_1 + A_2 + \cdots + A_n) = \sum_{i=1}^{n} Pr(A_i) - \sum_{j>i} Pr(A_i A_j)$$

$$+ \sum_{k>j>i} Pr(A_i A_j A_k)$$

$$- \sum_{l>k>j>i} Pr(A_i A_j A_k A_l) + \cdots \tag{1.3.10}$$

Rule 4a. For n nonindependent events $A_1, A_2, \ldots, A_n$,

$$\Pr(A_1 A_2 \cdots A_n) = \Pr(A_1)\Pr(A_2|A_1)\Pr(A_3|A_1 A_2) \cdots \quad (1.3.11)$$

Rule 4b. If the occurrence or nonoccurrence of event B depends on a previous event that can occur in n different ways, $A_1, A_2, \ldots, A_n$, then

$$\Pr(B) = \sum_{i=1}^{n} \Pr(B|A_i)\Pr(A_i) \quad (1.3.12)$$

where the A_i's are mutually exclusive events and hence Eq. 1.3.7 applies. Another important law, known as Bayes' formula, is derivable from Eq. 1.3.12.

Rule 4c (Bayes' formula).

$$\Pr(A_i|B) = \frac{\Pr(A_i)\Pr(B|A_i)}{\sum_{i=1}^{n}\Pr(B|A_i)\Pr(A_i)} \quad (1.3.13)$$

EXAMPLE 1.G

The third part of the example following Rule 4 is solved by applying Bayes' formula, Rule 4c. Make the identification

$$A_1 = A$$

$$A_2 = \overline{A}$$

Then,

$$\Pr(A|B) = \frac{\Pr(A)\Pr(B|A)}{\Pr(B|A) \cdot \Pr(A) + \Pr(B|A) \cdot \Pr(\overline{A})}$$

$$= \frac{(0.4)(0.25)}{(0.25)(0.4) + (0.5)(0.6)} = 0.25 \quad \blacksquare$$

1.4 RANDOM VARIABLE AND PROBABILITY DENSITY FUNCTIONS

The discussion of probability introduced the concept of the occurrence of some event (outcome). It is convenient to assign numbers to the possible outcomes, a process that often occurs quite naturally. For example, the outcome may be the measured percentage of iron in a sample, the Dow-Jones stock market average at a given time, the amount of uranium in a container of waste materials, or the observed total rainfall over a one-week period at a specified location.

Whenever a number is not assigned naturally, it can be assigned arbitrarily. For example, in inspecting items for defects, an item is classified as defective or nondefective. (The classification, of course, is not done arbitrarily.) These outcomes may be conveniently assigned the values zero and one, respectively, or one and zero if preferred.

Assigning numbers to experimental outcomes is accomplished by defining a *random variable*. A random variable is a numerically valued function defined over the elements of a sample space. To illustrate, if 10 items are inspected, with each one classified as either defective or nondefective, the random variable may be defined as the number of defective items among the 10 items. The possible values of the random variable are then $0, 1, 2, \ldots, 10$. As another example, for a moving production stream of items, the random variable may be defined as the number of items inspected before the fourth defect is found, in which case the possible values are $4, 5, 6, \ldots$.

1.4.1 Discrete and Continuous Random Variables

Random variables may be either discrete or continuous. When a sample space contains either a finite number or a countable infinity of elements, the random variable defined over that space is said to be *discrete*. When the random variable may take on any value in one or more intervals, it is said to be *continuous*. It is convenient to relate discrete random variables to counted data and continuous random variables to measured data.

Beginning with Chapter 2, we deal exclusively with continuous random variables involving measured data.

1.4.2 Probability Density Functions

The concepts of probability and random variables are united through a *probability density function* (pdf). A pdf is a function that associates a probability with each outcome or element of the sample space, that is, with each possible value that the random variable may assume. If x denotes a value of the random variable, $f(x)$ may denote the pdf.

The pdf may be expressed in tabular form, as indicated here:

x	$f(x)$
x_1	$f(x_1)$
x_2	$f(x_2)$
x_3	$f(x_3)$
$\vdots$	$\vdots$

It is usually more convenient to express a pdf as a mathematical function, noting that for a continuous random variable, the probability is $f(x)\,dx$. Since $f(x)$ in the discrete case and $f(x)\,dx$ in the continuous case are probabilities, it follows from the probability laws that $f(x)$ lies between 0 and 1 inclusive. For the discrete case,

$$\sum_i f(x_i) = 1 \qquad (1.4.1)$$

where the summation is over all possible values of the random variable. For the continuous case, summation is replaced by integration, so that the continuous-case equivalent of Eq. 1.4.1 is

$$\int_{-\infty}^{\infty} f(x)\,dx = 1 \qquad (1.4.2)$$

A distinction is made between a pdf and a cumulative distribution function (cdf). The cdf gives the probability that the random variable takes on a value less than or equal to x; the common notation is to replace the lowercase functional notation (f) by the corresponding capital letter (F). Thus for the discrete case,

$$F(x_i) = \Pr(x \le x_i) = \sum_{x \le x_i} f(x_i) \qquad (1.4.3)$$

For the continuous case,

$$F(x_i) = \Pr(x < x_i) = \int_{-\infty}^{x_i} f(x)\,dx \qquad (1.4.4)$$

Some specific pdf's are discussed briefly in Section 1.5. First, however, pdf's are characterized by their population moments.

1.4.3   Expectation (Moments)

Two descriptive values are useful in characterizing the pdf's of random variables, one a measure of central tendency and the other a measure of spread or dispersion. Since this book focuses on continuous pdf's, these measures are defined for such pdf's; the corresponding definitions for discrete pdf's are straightforward, the summation symbol replacing the integration symbol and dx being deleted.

Central tendency is conveniently described by the first moment about the origin, also called the expected value of the random variable or, more simply, the *mean*. Generally, the Greek letter μ is used to denote the mean:

$$\mu = E(x) = \int_{-\infty}^{\infty} xf(x)\, dx \qquad (1.4.5)$$

Borrowing from the terminology of mechanics, the mean is called the first moment about the origin because it is, in a sense, the center of gravity, the point around which the sum of distances to the left times the corresponding probabilities of occurrence balances out the same sum to the right.

Continuing with the terminology of mechanics, the second moment about the mean is a common descriptive measure of dispersion, called the *variance*. This quantity is generally denoted by the square of the Greek letter σ. The square root of σ^2 is called the *standard deviation*.

$$\sigma^2 = E(x - \mu)^2 = \int_{-\infty}^{\infty} (x - \mu)^2 f(x)\, dx \qquad (1.4.6)$$

The specific pdf that forms the basis for many of the results of this book, the normal pdf (see the discussion of normal pdf in Section 1.5.1), is completely characterized once μ and σ^2 are known. This is not true of all pdf's. Some pdf's are completely characterized by a single parameter. Further characterization of some pdf's requires knowledge of the third and fourth moments about the mean. The third moment is a measure of the skewness of the pdf, and the fourth moment is a measure of the peakedness, or kurtosis. Higher moments may also be defined, but in most cases, a knowledge of the first four moments supplies almost all the necessary information about the pdf.

The notation for the kth moment about the mean is as follows:

$$\mu_k = E(x - \mu)^k = \int_{-\infty}^{\infty} (x - \mu)^k f(x)\, dx \qquad (1.4.7)$$

In order to compare the skewness and peakedness of pdf's whose scales of measurement differ, the third and fourth moments are standardized. The standardized third moment is often denoted by $\sqrt{\beta_1}$,

$$\sqrt{\beta_1} = \frac{\mu_3}{\mu_2^{1.5}} \qquad (1.4.8)$$

and the standardized fourth moment by β_2,

$$\beta_2 = \frac{\mu_4}{\mu_2^2} \qquad (1.4.9)$$

We now consider some specific pdf's.

1.5 SPECIFIC PROBABILITY DENSITY FUNCTIONS

1.5.1 Univariate Probability Density Functions

A number of pdf's are useful, depending on the application. Many of the results in this book are based on the normal pdf and its multivariate counterpart, the multivariate normal pdf.

Univariate Normal pdf. The pdf is a two-parameter function of the form

$$f(x) = \frac{1}{\sqrt{2\pi}\, b} \exp\left[\frac{-(x-a)^2}{2b^2}\right] \tag{1.5.1}$$

The mean and standard deviation of this pdf are, respectively, a and b. Since the Greek letters μ and σ are generally used to denote the mean and standard deviation, respectively, the normal pdf is generally written in the form

$$f(x) = \frac{1}{\sqrt{2\pi}\, \sigma} \exp\left[\frac{-(x-\mu)^2}{2\sigma^2}\right] \tag{1.5.2}$$

The standardized third and fourth moments of this pdf are 0 and 3, respectively. The normal pdf is completely specified by the two parameters μ and σ.

A sketch of the normal pdf is given as Figure 1.1. The range of the random variable is, in theory, unbounded on both ends, but the probability that the random variable takes on a value more than three standard deviations in either direction from the mean is only 0.0027. Thus the normal pdf may also be used when a random variable can physically take on only positive values, even though in theory it can assume negative values as well.

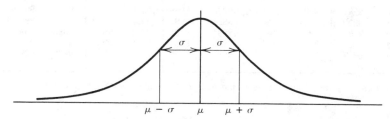

Figure 1.1   Graph of normal pdf.

Table 1.1 Cumulative Normal Distribution—Values of P^a

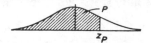

Values of P corresponding to z_p for the normal curve.

z is the standard normal variable. The value of P for $-z_p$ equals one minus the value of P for $+z_p$, e.g., the P for -1.62 equals $1 - .9474 = .0526$.

z_p	.00	.01	.02	.03	.04	.05	.06	.07	.08	.09
.0	.5000	.5040	.5080	.5120	.5160	.5199	.5239	.5279	.5319	.5359
.1	.5398	.5438	.5478	.5517	.5557	.5596	.5636	.5675	.5714	.5753
.2	.5793	.5832	.5871	.5910	.5948	.5987	.6026	.6064	.6103	.6141
.3	.6179	.6217	.6255	.6293	.6331	.6368	.6406	.6443	.6480	.6517
.4	.6554	.6591	.6628	.6664	.6700	.6736	.6772	.6808	.6844	.6879
.5	.6915	.6950	.6985	.7019	.7054	.7088	.7123	.7157	.7190	.7224
.6	.7257	.7291	.7324	.7357	.7389	.7422	.7454	.7486	.7517	.7549
.7	.7580	.7611	.7642	.7673	.7704	.7734	.7764	.7794	.7823	.7852
.8	.7881	.7910	.7939	.7967	.7995	.8023	.8051	.8078	.8106	.8133
.9	.8159	.8186	.8212	.8238	.8264	.8289	.8315	.8340	.8365	.8389
1.0	.8413	.8438	.8461	.8485	.8508	.8531	.8554	.8577	.8599	.8621
1.1	.8643	.8665	.8686	.8708	.8729	.8749	.8770	.8790	.8810	.8830
1.2	.8849	.8869	.8888	.8907	.8925	.8944	.8962	.8980	.8997	.9015
1.3	.9032	.9049	.9066	.9082	.9099	.9115	.9131	.9147	.9162	.9177
1.4	.9192	.9207	.9222	.9236	.9251	.9265	.9279	.9292	.9306	.9319
1.5	.9332	.9345	.9357	.9370	.9382	.9394	.9406	.9418	.9429	.9441
1.6	.9452	.9463	.9474	.9484	.9495	.9505	.9515	.9525	.9535	.9545
1.7	.9554	.9564	.9573	.9582	.9591	.9599	.9608	.9616	.9625	.9633
1.8	.9641	.9649	.9656	.9664	.9671	.9678	.9686	.9693	.9699	.9706
1.9	.9713	.9719	.9726	.9732	.9738	.9744	.9750	.9756	.9761	.9767
2.0	.9772	.9778	.9783	.9788	.9793	.9798	.9803	.9808	.9812	.9817
2.1	.9821	.9826	.9830	.9834	.9838	.9842	.9846	.9850	.9854	.9857
2.2	.9861	.9864	.9868	.9871	.9875	.9878	.9881	.9884	.9887	.9890
2.3	.9893	.9896	.9898	.9901	.9904	.9906	.9909	.9911	.9913	.9916
2.4	.9918	.9920	.9922	.9925	.9927	.9929	.9931	.9932	.9934	.9936
2.5	.9938	.9940	.9941	.9943	.9945	.9946	.9948	.9949	.9951	.9952
2.6	.9953	.9955	.9956	.9957	.9959	.9960	.9961	.9962	.9963	.9964
2.7	.9965	.9966	.9967	.9968	.9969	.9970	.9971	.9972	.9973	.9974
2.8	.9974	.9975	.9976	.9977	.9977	.9978	.9979	.9979	.9980	.9981
2.9	.9981	.9982	.9982	.9983	.9984	.9984	.9985	.9985	.9986	.9986
3.0	.9987	.9987	.9987	.9988	.9988	.9989	.9989	.9989	.9990	.9990
3.1	.9990	.9991	.9991	.9991	.9992	.9992	.9992	.9992	.9993	.9993
3.2	.9993	.9993	.9994	.9994	.9994	.9994	.9994	.9995	.9995	.9995
3.3	.9995	.9995	.9995	.9996	.9996	.9996	.9996	.9996	.9996	.9997
3.4	.9997	.9997	.9997	.9997	.9997	.9997	.9997	.9997	.9997	.9998

^{a}Reprinted from the United States Department of Commerce NBS Handbook 91.

From Eqs. 1.4.4 and 1.5.2, the cdf for the normal pdf is

$$F(x_i) = \frac{1}{\sqrt{2\pi}\,\sigma} \int_{-\infty}^{x_i} \exp\left[\frac{-(x-\mu)^2}{2\sigma^2}\right] dx \qquad (1.5.3)$$

This is not integrable in closed form, and numerical methods would be required to compute $F(x_i)$ for given μ, σ, and x_i if tables did not exist. Clearly, one could not provide tables for all combinations of μ and σ, but such tables are not necessary. If the following transformation is made:

$$z = \frac{x-\mu}{\sigma} \qquad (1.5.4)$$

then the pdf of z is the normal pdf with mean zero and standard deviation one. The standardized normal cdf is widely tabulated and is included here as Table 1.1.

EXAMPLE 1.H

To illustrate, assume that the random variable X is the gross weight of a container of some product, and that x is normally distributed with mean $\mu = 19.84$ ounces and standard deviation $\sigma = 0.24$ ounce. Then the probability that a container selected at random will weigh less than 20.20 ounces is written

$$Pr(x < 20.20 | \mu = 19.84; \sigma = 0.24)$$

In applying the transformation in Eq. 1.5.4, this probability may be rewritten

$$Pr\left(\frac{x - 19.84}{0.24} < \frac{20.20 - 19.84}{0.24}\right) = Pr(z < 1.5)$$

In Table 1.1, z_p is 1.5, and so the required probability is 0.9332.

If the problem were to find the probability that the container selected at random weighed more than 20.20 ounces, this probability would follow immediately from Eq. 1.3.2 and would be $1 - 0.9332 = 0.0668$. ∎

In finding the probability that the random variable takes on a value between two specific numbers, two entries from Table 1.1 are required. The sketch at the top of the table, along with the symmetry of the normal pdf, is helpful, as the following example illustrates.

EXAMPLE 1.1

Find the probability that the gross container weight is between 19.60 and 20.20 ounces. The solution is

$$Pr(19.60 < x < 20.20) = Pr\left(\frac{19.60 - 19.84}{0.24} < z < \frac{20.20 - 19.84}{0.24}\right)$$

$$= Pr(-1.0 < z < 1.5)$$

$$= 0.9332 - (1 - 0.8413) = 0.7745$$

Figure 1.2 is helpful in understanding the last step in the above calculations.

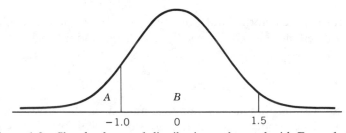

Figure 1.2   Sketch of normal distribution to be used with Example 1.1.

$$\text{required probability} = B$$

$$A + B = 0.9332$$

$$A = 1 - 0.8413$$

$$B = (A + B) - A \qquad ■$$

Uniform pdf.   Although the uniform pdf is rarely mentioned in later chapters, it is discussed briefly to provide a complete discussion of errors of measurement. In some applications, errors due to rounding may be major contributors to total errors of measurement, and the rounding error is a random variable with uniform pdf.

The uniform pdf is of the form

$$f(x) = \frac{1}{b - a}, \qquad a < x < b$$

$$= 0, \qquad \text{elsewhere} \qquad (1.5.5)$$

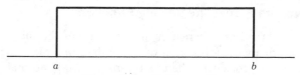

Figure 1.3   Uniform pdf.

This pdf is sketched in Figure 1.3; it is a rectangle with height $1/(b - a)$ and width $(b - a)$, and the area of the rectangle clearly equals unity.

By direct application of Eqs. 1.4.5 and 1.4.6, respectively, it is easy to show that for the uniform pdf,

$$\mu = \frac{a + b}{2} \qquad (1.5.6)$$

$$\sigma^2 = \frac{(b - a)^2}{12} \qquad (1.5.7)$$

The value of σ^2 is of special interest, for it gives the variance due to rounding in a measurement process, as the following example illustrates.

EXAMPLE 1.J

In analyzing a uranium dioxide compound for hydrogen content, the result of an analysis is reported to the nearest 0.1 ppm. Therefore, from Eq. 1.5.7, the rounding error variance is

$$\sigma^2 = \frac{0.01}{12} = 8.333 \times 10^{-4} \text{ ppm}^2$$

and the standard deviation is 0.0287 ppm.   ∎

For the uniform pdf, the corresponding cdf is easily found by direct integration of $f(x)$ in Eq. 1.5.5.

Other pdf's.   The normal and uniform univariate pdf's are not the only two pdf's. They are the important pdf's when dealing with errors of measurement; in other applications, other pdf's would become important. However, space does not permit even a brief discussion of additional pdf's, both continuous and discrete. For an excellent summary of common and not-so-common pdf's, the reader is referred to Hahn and Shapiro (2), Figures 3.21 and 4.8 and Tables 3.2 and 4.3.

1.5.2   Multivariate Probability Density Functions

Beginning in Chapter 3 and thereafter, use is made of the *multivariate normal pdf*, which is a pdf that is a function of more than one random variable. The only multivariate pdf considered here is the multivariate normal.

One particular multivariate normal pdf is the *bivariate* normal pdf, involving just two random variables. The bivariate normal pdf forms the basis for some key results in Chapter 3.

The bivariate pdf with X_1 and X_2 as the random variables may be written

$$f(x_1, x_2) = \frac{1}{2\pi\sigma_1\sigma_2\sqrt{1 - \rho^2}}$$

$$\times \exp\left\{-\frac{1}{2(1 - \rho^2)}\left[\frac{(x_1 - \mu_1)^2}{\sigma_1^2} - \frac{2\rho(x_1 - \mu_1)(x_2 - \mu_2)}{\sigma_1\sigma_2}\right.\right.$$

$$\left.\left.+\frac{(x_2 - \mu_2)^2}{\sigma_2^2}\right]\right\} \tag{1.5.8}$$

where the parameters are μ_1, μ_2, σ_1^2, σ_2^2, and ρ, with

$$E(x_i) = \mu_i; \qquad i = 1, 2 \tag{1.5.9}$$

$$E(x_i - \mu_i)^2 = \sigma_i^2; \qquad i = 1, 2 \tag{1.5.10}$$

$$\sigma_{12} = \sigma_{21} = E\left[(x_1 - \mu_1)(x_2 - \mu_2)\right] = \rho\sigma_1\sigma_2 \tag{1.5.11}$$

The quantity σ_{12} or σ_{21} defined by Eq. 1.5.11 is called the *covariance* of x_1 and x_2, and ρ is called the *correlation coefficient*. When $\rho = 0$, $f(x_1, x_2)$ is simply the product of two univariate pdf's, and x_1 and x_2 are said to be *independently distributed normal random variables*.

In extending the bivariate normal pdf to include $k > 2$ random variables, $X_1, X_2, \ldots, X_k$, the multivariate pdf is written

$$f(x_1, x_2, \ldots, x_k) = \left(\frac{1}{2\pi}\right)^{0.5k}\sqrt{|\sigma^{ij}|}\exp\left[-\frac{1}{2}\sum_{i=1}^{k}\sum_{j=1}^{k}\sigma^{ij}(x_i - \mu_i)(x_j - \mu_j)\right]$$

$$\tag{1.5.12}$$

where the symmetric matrix (σ_{ij}) is called the *variance-covariance matrix*, which has element σ_i^2 in row i, column i and element σ_{ij} in row i, column j.

The determinant denoted by $|\sigma^{ij}|$ in Eq. 1.5.12 is the reciprocal of the determinant of $|\sigma_{ij}|$. The quantity σ^{ij} in Eq. 1.5.12 is the element in row i, column j of the inverse matrix:

$$\left(\sigma^{ij} \right) = \left(\sigma_{ij} \right)^{-1} \tag{1.5.13}$$

Later chapters discuss and give examples of the application of the multivariate normal pdf. Matrix algebra is presented briefly in Section 1.9.

1.6 SAMPLING IN STATISTICAL INFERENCE

1.6.1 Populations

In statistics, a *population* is any complete collection of objects, items, or the results of operations. In this book, a population usually is the conceptually complete collection of the results of a measurement operation. A population may be finite in size or of infinite size, a distinction that is of some importance in later chapters. From a statistics viewpoint, an experiment is conducted to obtain some specified information about some population. It is often impossible or impracticable to measure (in some sense) the entire population, and so one limits this measurement process to some part or sample of the population. Inferences are then made about the entire population on the basis of this sample.

1.6.2 The Sampling Process

An important but sometimes overlooked requirement is that the sample must be representative of the population to be studied. Where sample information has not led to correct inferences about a population, the reason is often related to inadequacies in drawing the sample. For example, a radio station may conduct a listeners' poll on some subject of current interest. Such a sample can scarcely be called representative, for only those individuals who have strong viewpoints on the subject tend to volunteer.

The quality of representativeness is often minimized with respect to its importance. To illustrate, a laboratory manager who wishes to characterize a measurement process by measuring a given characteristic of some sample material may do so by having one analyst perform all measurements on one of perhaps many nominally identical instruments with all measurements performed within, say, a two-hour time span. It is clear that the result is only a narrow view of the measurement process as a whole. Differences among analysts, instruments, or time-related factors are not investigated in this experiment. However, such an experiment is valid if the inferences are limited to the restricted population, possibly to evaluate the ultimate capability of the measurement process under fixed conditions. The important lesson is that it is necessary to keep the population in mind when making inferences on the basis of the sample results.

An oft-cited requirement of a sample is that the items in the sample be drawn randomly, where the randomness refers to the method of drawing the sample and not to the sample results. For example, one could draw a random sample of size 20 from the population of all residents of the city of Seattle and have all sample members be male. Such an occurrence is quite unlikely, and *after-the-fact* might be labeled a nonrandom sample. Nevertheless, the process of drawing the sample may have been truly random. It is this quandary that may lead to a certain amount of nonrandomness in the sample selection, coupled with randomness. In the example just considered, to insure an even distribution of males and females, one could divide the population of residents into these two identified subpopulations, then draw the individuals randomly within the subpopulations.

In the specific experimental situations to be considered later in this book, the populations are carefully defined, as is the process of drawing the samples.

1.6.3 Sampling Distributions

In statistical inference, a *statistic* is a function of observations and is hence a random variable with its own pdf. The bases for employing certain statistics are presented in Section 1.7. For immediate purposes, we define the statistics in question and consider their pdf's.

Let $x_1, x_2, \ldots, x_n$ denote the sample observations selected at random from some population. The two sample statistics of primary interest are the sample mean $\bar{x}$, often called simply the *average*, and the sample variance, s^2, defined as follows:

$$\bar{x} = \sum_{i=1}^{n} \frac{x_i}{n} \tag{1.6.1}$$

$$s^2 = \frac{\sum_{i=1}^{n} x_i^2 - \left(\sum_{i=1}^{n} x_i\right)^2 / n}{n-1} \tag{1.6.2}$$

The pdf's for these statistics and for special functions of them are considered in the following sections, in which it is assumed that $x_1, x_2, \ldots, x_n$ are randomly selected from a normal pdf with mean μ and variance σ^2.

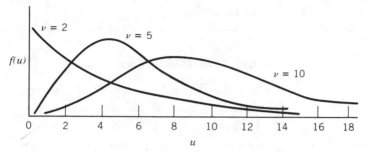

Figure 1.4 Graphs of chi-square pdf's.

Chi-Square. The random variable

$$u = \frac{(n-1)s^2}{\sigma^2} \tag{1.6.3}$$

has the chi-square pdf with parameter $(n-1)$ called the *degrees of freedom*. In general, the chi-square pdf is of the form

$$f(u) = \chi^2_{(\nu)} = C_1 u^{(\nu-2)/2} \exp\left(\frac{-u}{2}\right), \qquad u \geq 0 \tag{1.6.4}$$

where the parameter ν is called the degrees of freedom, and C_1 is a known function of ν. Graphs of the chi-square pdf for a few values of ν are shown in Figure 1.4.

Since the chi-square pdf is a single parameter pdf, it is simple to tabulate. A table of the cdf is given as Table 1.2. In that table, $\chi^2_{P(\nu)}$ is given as a function of P and degrees of freedom, $df = \nu$, where $\chi^2_{P(\nu)}$ is defined by

$$\int_0^{\chi^2_{P(\nu)}} \chi^2_{(\nu)} \, d\chi^2 = P \tag{1.6.5}$$

EXAMPLE 1.K

If $n = 5$ and $\sigma^2 = 4$ units^2, find the probability that s^2 exceeds 7.78. Here $\nu = (n-1) = 4$ so that

$$\Pr(s^2 > 7.78) = \Pr\left[u > \frac{4(7.78)}{4}\right]$$

$$= \Pr(u > 7.78)$$

$$= 0.10 \text{ from Table 1.2} \qquad \blacksquare$$

Table 1.2   Percentiles of the χ^2 Distribution[a]

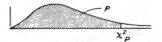

Values of χ_P^2 corresponding to P

df	$\chi^2_{.005}$	$\chi^2_{.01}$	$\chi^2_{.025}$	$\chi^2_{.05}$	$\chi^2_{.10}$	$\chi^2_{.90}$	$\chi^2_{.95}$	$\chi^2_{.975}$	$\chi^2_{.99}$	$\chi^2_{.995}$
1	.000039	.00016	.00098	.0039	.0158	2.71	3.84	5.02	6.63	7.88
2	.0100	.0201	.0506	.1026	.2107	4.61	5.99	7.38	9.21	10.60
3	.0717	.115	.216	.352	.584	6.25	7.81	9.35	11.34	12.84
4	.207	.297	.484	.711	1.064	7.78	9.49	11.14	13.28	14.86
5	.412	.554	.831	1.15	1.61	9.24	11.07	12.83	15.09	16.75
6	.676	.872	1.24	1.64	2.20	10.64	12.59	14.45	16.81	18.55
7	.989	1.24	1.69	2.17	2.83	12.02	14.07	16.01	18.48	20.28
8	1.34	1.65	2.18	2.73	3.49	13.36	15.51	17.53	20.09	21.96
9	1.73	2.09	2.70	3.33	4.17	14.68	16.92	19.02	21.67	23.59
10	2.16	2.56	3.25	3.94	4.87	15.99	18.31	20.48	23.21	25.19
11	2.60	3.05	3.82	4.57	5.58	17.28	19.68	21.92	24.73	26.76
12	3.07	3.57	4.40	5.23	6.30	18.55	21.03	23.34	26.22	28.30
13	3.57	4.11	5.01	5.89	7.04	19.81	22.36	24.74	27.69	29.82
14	4.07	4.66	5.63	6.57	7.79	21.06	23.68	26.12	29.14	31.32
15	4.60	5.23	6.26	7.26	8.55	22.31	25.00	27.49	30.58	32.80
16	5.14	5.81	6.91	7.96	9.31	23.54	26.30	28.85	32.00	34.27
18	6.26	7.01	8.23	9.39	10.86	25.99	28.87	31.53	34.81	37.16
20	7.43	8.26	9.59	10.85	12.44	28.41	31.41	34.17	37.57	40.00
24	9.89	10.86	12.40	13.85	15.66	33.20	36.42	39.36	42.98	45.56
30	13.79	14.95	16.79	18.49	20.60	40.26	43.77	46.98	50.89	53.67
40	20.71	22.16	24.43	26.51	29.05	51.81	55.76	59.34	63.69	66.77
60	35.53	37.48	40.48	43.19	46.46	74.40	79.08	83.30	88.38	91.95
120	83.85	86.92	91.58	95.70	100.62	140.23	146.57	152.21	158.95	163.64

[a] Reprinted from the United States Department of Commerce NBS Handbook 91.

Student's t.   With $\bar{x}$ defined by Eq. 1.6.1, it is a random variable with normal pdf. Its mean is μ, and its variance is σ^2/n. By Eq. 1.5.4, the random variable

$$z = \frac{\bar{x} - \mu}{\sigma/\sqrt{n}} \qquad (1.6.6)$$

is normally distributed with zero mean and unit variance.

If σ were known, inferences could be made about μ given $\bar{x}$ and the pdf of the statistic defined by Eq. 1.6.6. In most cases, σ is not known and is replaced

Table 1.3 Percentiles of the t Distribution[a]

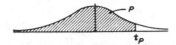

df	$t_{.60}$	$t_{.70}$	$t_{.80}$	$t_{.90}$	$t_{.95}$	$t_{.975}$	$t_{.99}$	$t_{.995}$
1	.325	.727	1.376	3.078	6.314	12.706	31.821	63.657
2	.289	.617	1.061	1.886	2.920	4.303	6.965	9.925
3	.277	.584	.978	1.638	2.353	3.182	4.541	5.841
4	.271	.569	.941	1.533	2.132	2.776	3.747	4.604
5	.267	.559	.920	1.476	2.015	2.571	3.365	4.032
6	.265	.553	.906	1.440	1.943	2.447	3.143	3.707
7	.263	.549	.896	1.415	1.895	2.365	2.998	3.499
8	.262	.546	.889	1.397	1.860	2.306	2.896	3.355
9	.261	.543	.883	1.383	1.833	2.262	2.821	3.250
10	.260	.542	.879	1.372	1.812	2.228	2.764	3.169
11	.260	.540	.876	1.363	1.796	2.201	2.718	3.106
12	.259	.539	.873	1.356	1.782	2.179	2.681	3.055
13	.259	.538	.870	1.350	1.771	2.160	2.650	3.012
14	.258	.537	.868	1.345	1.761	2.145	2.624	2.977
15	.258	.536	.866	1.341	1.753	2.131	2.602	2.947
16	.258	.535	.865	1.337	1.746	2.120	2.583	2.921
17	.257	.534	.863	1.333	1.740	2.110	2.567	2.898
18	.257	.534	.862	1.330	1.734	2.101	2.552	2.878
19	.257	.533	.861	1.328	1.729	2.093	2.539	2.861
20	.257	.533	.860	1.325	1.725	2.086	2.528	2.845
21	.257	.532	.859	1.323	1.721	2.080	2.518	2.831
22	.256	.532	.858	1.321	1.717	2.074	2.508	2.819
23	.256	.532	.858	1.319	1.714	2.069	2.500	2.807
24	.256	.531	.857	1.318	1.711	2.064	2.492	2.797
25	.256	.531	.856	1.316	1.708	2.060	2.485	2.787
26	.256	.531	.856	1.315	1.706	2.056	2.479	2.779
27	.256	.531	.855	1.314	1.703	2.052	2.473	2.771
28	.256	.530	.855	1.313	1.701	2.048	2.467	2.763
29	.256	.530	.854	1.311	1.699	2.045	2.462	2.756
30	.256	.530	.854	1.310	1.697	2.042	2.457	2.750
40	.255	.529	.851	1.303	1.684	2.021	2.423	2.704
60	.254	.527	.848	1.296	1.671	2.000	2.390	2.660
120	.254	.526	.845	1.289	1.658	1.980	2.358	2.617
∞	.253	.524	.842	1.282	1.645	1.960	2.326	2.576

[a] Reprinted from the United States Department of Commerce NBS Handbook 91.

by an estimate, s, with s^2 as defined in Eq. 1.6.2. The statistic then becomes

$$t = \frac{\bar{x} - \mu}{s/\sqrt{n}} \tag{1.6.7}$$

This statistic no longer has the normal pdf with zero mean and unit variance. Intuitively, one would expect the pdf of t to be broader than the standardized normal pdf because the constant σ has now been replaced by the random variable s. The pdf of t is known as Student's t, with parameter $(n - 1)$, called the degrees of freedom. In general, the t pdf is of the form

$$f(t) = C_2 \left[\frac{\nu + t^2}{\nu} \right]^{-(\nu+1)/2}, \qquad -\infty < t < \infty \tag{1.6.8}$$

where C_2 is a known function of ν, and where the parameter ν is called the degrees of freedom. The t pdf, being a single-parameter pdf, is simple to tabulate. A table of the cdf is given as Table 1.3, wherein $t_{P(\nu)}$ is given as a function of P for fixed parameter value ν, with $t_{P(\nu)}$ defined by

$$\int_{-\infty}^{t_{P(\nu)}} f(t)\, dt = P \tag{1.6.9}$$

The symmetry of the t pdf permits tabulation of the cdf for values of $P > 0.5$. The t pdf is symmetrical about $t = 0$, with a lower peak and broader tails than a normal pdf. As the number of degrees of freedom increases, the t pdf approaches the standardized normal pdf.

EXAMPLE 1.L

To illustrate the use of Table 1.3, if $n = 9$, $\bar{x} = 3$, and $s = 4$, the following probability statement applies:

$$\Pr\left(\frac{3 - \mu}{4/3} > 0.889 \right) = 0.20$$

Solving for μ gives

$$\Pr(\mu > 1.815) = 0.80$$

This last probability statement provides what is known as a one-sided confidence interval on the parameter μ (see Section 1.7.3 for a further discussion on confidence intervals). In the above equation, μ is a constant while 1.815 is a realization of the random variable that is a function of $\bar{x}$ and s. ∎

Table 1.4 Percentiles of the F Distribution[a]

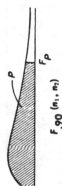

$F_{.90}\,(n_1, n_2)$

n_1 = degrees of freedom for numerator

n_2 = degrees of freedom for denominator

n_2 \ n_1	1	2	3	4	5	6	7	8	9	10	12	15	20	24	30	40	60	120	∞
1	39.86	49.50	53.59	55.83	57.24	58.20	58.91	59.44	59.86	60.19	60.71	61.22	61.74	62.00	62.26	62.53	62.79	63.06	63.33
2	8.53	9.00	9.16	9.24	9.29	9.33	9.35	9.37	9.38	9.39	9.41	9.42	9.44	9.45	9.46	9.47	9.47	9.48	9.49
3	5.54	5.46	5.39	5.34	5.31	5.28	5.27	5.25	5.24	5.23	5.22	5.20	5.18	5.18	5.17	5.16	5.15	5.14	5.13
4	4.54	4.32	4.19	4.11	4.05	4.01	3.98	3.95	3.94	3.92	3.90	3.87	3.84	3.83	3.82	3.80	3.79	3.78	3.76
5	4.06	3.78	3.62	3.52	3.45	3.40	3.37	3.34	3.32	3.30	3.27	3.24	3.21	3.19	3.17	3.16	3.14	3.12	3.10
6	3.78	3.46	3.29	3.18	3.11	3.05	3.01	2.98	2.96	2.94	2.90	2.87	2.84	2.82	2.80	2.78	2.76	2.74	2.72
7	3.59	3.26	3.07	2.96	2.88	2.83	2.78	2.75	2.72	2.70	2.67	2.63	2.59	2.58	2.56	2.54	2.51	2.49	2.47
8	3.46	3.11	2.92	2.81	2.73	2.67	2.62	2.59	2.56	2.54	2.50	2.46	2.42	2.40	2.38	2.36	2.34	2.32	2.29
9	3.36	3.01	2.81	2.69	2.61	2.55	2.51	2.47	2.44	2.42	2.38	2.34	2.30	2.28	2.25	2.23	2.21	2.18	2.16
10	3.29	2.92	2.73	2.61	2.52	2.46	2.41	2.38	2.35	2.32	2.28	2.24	2.20	2.18	2.16	2.13	2.11	2.08	2.06
11	3.23	2.86	2.66	2.54	2.45	2.39	2.34	2.30	2.27	2.25	2.21	2.17	2.12	2.10	2.08	2.05	2.03	2.00	1.97
12	3.18	2.81	2.61	2.48	2.39	2.33	2.28	2.24	2.21	2.19	2.15	2.10	2.06	2.04	2.01	1.99	1.96	1.93	1.90
13	3.14	2.76	2.56	2.43	2.35	2.28	2.23	2.20	2.16	2.14	2.10	2.05	2.01	1.98	1.96	1.93	1.90	1.88	1.85
14	3.10	2.73	2.52	2.39	2.31	2.24	2.19	2.15	2.12	2.10	2.05	2.01	1.96	1.94	1.91	1.89	1.86	1.83	1.80
15	3.07	2.70	2.49	2.36	2.27	2.21	2.16	2.12	2.09	2.06	2.02	1.97	1.92	1.90	1.87	1.85	1.82	1.79	1.76
16	3.05	2.67	2.46	2.33	2.24	2.18	2.13	2.09	2.06	2.03	1.99	1.94	1.89	1.87	1.84	1.81	1.78	1.75	1.72
17	3.03	2.64	2.44	2.31	2.22	2.15	2.10	2.06	2.03	2.00	1.96	1.91	1.86	1.84	1.81	1.78	1.75	1.72	1.69
18	3.01	2.62	2.42	2.29	2.20	2.13	2.08	2.04	2.00	1.98	1.93	1.89	1.84	1.81	1.78	1.75	1.72	1.69	1.66
19	2.99	2.61	2.40	2.27	2.18	2.11	2.06	2.02	1.98	1.96	1.91	1.86	1.81	1.79	1.76	1.73	1.70	1.67	1.63
20	2.97	2.59	2.38	2.25	2.16	2.09	2.04	2.00	1.96	1.94	1.89	1.84	1.79	1.77	1.74	1.71	1.68	1.64	1.61
21	2.96	2.57	2.36	2.23	2.14	2.08	2.02	1.98	1.95	1.92	1.87	1.83	1.78	1.75	1.72	1.69	1.66	1.62	1.59
22	2.95	2.56	2.35	2.22	2.13	2.06	2.01	1.97	1.93	1.90	1.86	1.81	1.76	1.73	1.70	1.67	1.64	1.60	1.57
23	2.94	2.55	2.34	2.21	2.11	2.05	1.99	1.95	1.92	1.89	1.84	1.80	1.74	1.72	1.69	1.66	1.62	1.59	1.55
24	2.93	2.54	2.33	2.19	2.10	2.04	1.98	1.94	1.91	1.88	1.83	1.78	1.73	1.70	1.67	1.64	1.61	1.57	1.53
25	2.92	2.53	2.32	2.18	2.09	2.02	1.97	1.93	1.89	1.87	1.82	1.77	1.72	1.69	1.66	1.63	1.59	1.56	1.52
26	2.91	2.52	2.31	2.17	2.08	2.01	1.96	1.92	1.88	1.86	1.81	1.76	1.71	1.68	1.65	1.61	1.58	1.54	1.50
27	2.90	2.51	2.30	2.17	2.07	2.00	1.95	1.91	1.87	1.85	1.80	1.75	1.70	1.67	1.64	1.60	1.57	1.53	1.49
28	2.89	2.50	2.29	2.16	2.06	2.00	1.94	1.90	1.87	1.84	1.79	1.74	1.69	1.66	1.63	1.59	1.56	1.52	1.48
29	2.89	2.50	2.28	2.15	2.06	1.99	1.93	1.89	1.86	1.83	1.78	1.73	1.68	1.65	1.62	1.58	1.55	1.51	1.47
30	2.88	2.49	2.28	2.14	2.05	1.98	1.93	1.88	1.85	1.82	1.77	1.72	1.67	1.64	1.61	1.57	1.54	1.50	1.46
40	2.84	2.44	2.23	2.09	2.00	1.93	1.87	1.83	1.79	1.76	1.71	1.66	1.61	1.57	1.54	1.51	1.47	1.42	1.38
60	2.79	2.39	2.18	2.04	1.95	1.87	1.82	1.77	1.74	1.71	1.66	1.60	1.54	1.51	1.48	1.44	1.40	1.35	1.29
120	2.75	2.35	2.13	1.99	1.90	1.82	1.77	1.72	1.68	1.65	1.60	1.55	1.48	1.45	1.41	1.37	1.32	1.26	1.19
∞	2.71	2.30	2.08	1.94	1.85	1.77	1.72	1.67	1.63	1.60	1.55	1.49	1.42	1.38	1.34	1.30	1.24	1.17	1.00

Table 1.4 (Continued)

$F_{.95}(n_1, n_2)$

n_1 = degrees of freedom for numerator

n_2 \ n_1	1	2	3	4	5	6	7	8	9	10	12	15	20	24	30	40	60	120	∞
1	161.4	199.5	215.7	224.6	230.2	234.0	236.8	238.9	240.5	241.9	243.9	245.9	248.0	249.1	250.1	251.1	252.2	253.3	254.3
2	18.51	19.00	19.16	19.25	19.30	19.33	19.35	19.37	19.38	19.40	19.41	19.43	19.45	19.45	19.46	19.47	19.48	19.49	19.50
3	10.13	9.55	9.28	9.12	9.01	8.94	8.89	8.85	8.81	8.79	8.74	8.70	8.66	8.64	8.62	8.59	8.57	8.55	8.53
4	7.71	6.94	6.59	6.39	6.26	6.16	6.09	6.04	6.00	5.96	5.91	5.86	5.80	5.77	5.75	5.72	5.69	5.66	5.63
5	6.61	5.79	5.41	5.19	5.05	4.95	4.88	4.82	4.77	4.74	4.68	4.62	4.56	4.53	4.50	4.46	4.43	4.40	4.36
6	5.99	5.14	4.76	4.53	4.39	4.28	4.21	4.15	4.10	4.06	4.00	3.94	3.87	3.84	3.81	3.77	3.74	3.70	3.67
7	5.59	4.74	4.35	4.12	3.97	3.87	3.79	3.73	3.68	3.64	3.57	3.51	3.44	3.41	3.38	3.34	3.30	3.27	3.23
8	5.32	4.46	4.07	3.84	3.69	3.58	3.50	3.44	3.39	3.35	3.28	3.22	3.15	3.12	3.08	3.04	3.01	2.97	2.93
9	5.12	4.26	3.86	3.63	3.48	3.37	3.29	3.23	3.18	3.14	3.07	3.01	2.94	2.90	2.86	2.83	2.79	2.75	2.71
10	4.96	4.10	3.71	3.48	3.33	3.22	3.14	3.07	3.02	2.98	2.91	2.85	2.77	2.74	2.70	2.66	2.62	2.58	2.54
11	4.84	3.98	3.59	3.36	3.20	3.09	3.01	2.95	2.90	2.85	2.79	2.72	2.65	2.61	2.57	2.53	2.49	2.45	2.40
12	4.75	3.89	3.49	3.26	3.11	3.00	2.91	2.85	2.80	2.75	2.69	2.62	2.54	2.51	2.47	2.43	2.38	2.34	2.30
13	4.67	3.81	3.41	3.18	3.03	2.92	2.83	2.77	2.71	2.67	2.60	2.53	2.46	2.42	2.38	2.34	2.30	2.25	2.21
14	4.60	3.74	3.34	3.11	2.96	2.85	2.76	2.70	2.65	2.60	2.53	2.46	2.39	2.35	2.31	2.27	2.22	2.18	2.13
15	4.54	3.68	3.29	3.06	2.90	2.79	2.71	2.64	2.59	2.54	2.48	2.40	2.33	2.29	2.25	2.20	2.16	2.11	2.07
16	4.49	3.63	3.24	3.01	2.85	2.74	2.66	2.59	2.54	2.49	2.42	2.35	2.28	2.24	2.19	2.15	2.11	2.06	2.01
17	4.45	3.59	3.20	2.96	2.81	2.70	2.61	2.55	2.49	2.45	2.38	2.31	2.23	2.19	2.15	2.10	2.06	2.01	1.96
18	4.41	3.55	3.16	2.93	2.77	2.66	2.58	2.51	2.46	2.41	2.34	2.27	2.19	2.15	2.11	2.06	2.02	1.97	1.92
19	4.38	3.52	3.13	2.90	2.74	2.63	2.54	2.48	2.42	2.38	2.31	2.23	2.16	2.11	2.07	2.03	1.98	1.93	1.88
20	4.35	3.49	3.10	2.87	2.71	2.60	2.51	2.45	2.39	2.35	2.28	2.20	2.12	2.08	2.04	1.99	1.95	1.90	1.84
21	4.32	3.47	3.07	2.84	2.68	2.57	2.49	2.42	2.37	2.32	2.25	2.18	2.10	2.05	2.01	1.96	1.92	1.87	1.81
22	4.30	3.44	3.05	2.82	2.66	2.55	2.46	2.40	2.34	2.30	2.23	2.15	2.07	2.03	1.98	1.94	1.89	1.84	1.78
23	4.28	3.42	3.03	2.80	2.64	2.53	2.44	2.37	2.32	2.27	2.20	2.13	2.05	2.01	1.96	1.91	1.86	1.81	1.76
24	4.26	3.40	3.01	2.78	2.62	2.51	2.42	2.36	2.30	2.25	2.18	2.11	2.03	1.98	1.94	1.89	1.84	1.79	1.73
25	4.24	3.39	2.99	2.76	2.60	2.49	2.40	2.34	2.28	2.24	2.16	2.09	2.01	1.96	1.92	1.87	1.82	1.77	1.71
26	4.23	3.37	2.98	2.74	2.59	2.47	2.39	2.32	2.27	2.22	2.15	2.07	1.99	1.95	1.90	1.85	1.80	1.75	1.69
27	4.21	3.35	2.96	2.73	2.57	2.46	2.37	2.31	2.25	2.20	2.13	2.06	1.97	1.93	1.88	1.84	1.79	1.73	1.67
28	4.20	3.34	2.95	2.71	2.56	2.45	2.36	2.29	2.24	2.19	2.12	2.04	1.96	1.91	1.87	1.82	1.77	1.71	1.65
29	4.18	3.33	2.93	2.70	2.55	2.43	2.35	2.28	2.22	2.18	2.10	2.03	1.94	1.90	1.85	1.81	1.75	1.70	1.64
30	4.17	3.32	2.92	2.69	2.53	2.42	2.33	2.27	2.21	2.16	2.09	2.01	1.93	1.89	1.84	1.79	1.74	1.68	1.62
40	4.08	3.23	2.84	2.61	2.45	2.34	2.25	2.18	2.12	2.08	2.00	1.92	1.84	1.79	1.74	1.69	1.64	1.58	1.51
60	4.00	3.15	2.76	2.53	2.37	2.25	2.17	2.10	2.04	1.99	1.92	1.84	1.75	1.70	1.65	1.59	1.53	1.47	1.39
120	3.92	3.07	2.68	2.45	2.29	2.17	2.09	2.02	1.96	1.91	1.83	1.75	1.66	1.61	1.55	1.50	1.43	1.35	1.25
∞	3.84	3.00	2.60	2.37	2.21	2.10	2.01	1.94	1.88	1.83	1.75	1.67	1.57	1.52	1.46	1.39	1.32	1.22	1.00

n_2 = degrees of freedom for denominator

F.975 (n₁, n₂)

$F_{.975}(n_1, n_2)$

n_1 = degrees of freedom for numerator

n_2	1	2	3	4	5	6	7	8	9	10	12	15	20	24	30	40	60	120	∞
1	647.8	799.5	864.2	899.6	921.8	937.1	948.2	956.7	963.3	968.6	976.7	984.9	993.1	997.2	1001	1006	1010	1014	1018
2	38.51	39.00	39.17	39.25	39.30	39.33	39.36	39.37	39.39	39.40	39.41	39.43	39.45	39.46	39.46	39.47	39.48	39.49	39.50
3	17.44	16.04	15.44	15.10	14.88	14.73	14.62	14.54	14.47	14.42	14.34	14.25	14.17	14.12	14.08	14.04	13.99	13.95	13.90
4	12.22	10.65	9.98	9.60	9.36	9.20	9.07	8.98	8.90	8.84	8.75	8.66	8.56	8.51	8.46	8.41	8.36	8.31	8.26
5	10.01	8.43	7.76	7.39	7.15	6.98	6.85	6.76	6.68	6.62	6.52	6.43	6.33	6.28	6.23	6.18	6.12	6.07	6.02
6	8.81	7.26	6.60	6.23	5.99	5.82	5.70	5.60	5.52	5.46	5.37	5.27	5.17	5.12	5.07	5.01	4.96	4.90	4.85
7	8.07	6.54	5.89	5.52	5.29	5.12	4.99	4.90	4.82	4.76	4.67	4.57	4.47	4.42	4.36	4.31	4.25	4.20	4.14
8	7.57	6.06	5.42	5.05	4.82	4.65	4.53	4.43	4.36	4.30	4.20	4.10	4.00	3.95	3.89	3.84	3.78	3.73	3.67
9	7.21	5.71	5.08	4.72	4.48	4.32	4.20	4.10	4.03	3.96	3.87	3.77	3.67	3.61	3.56	3.51	3.45	3.39	3.33
10	6.94	5.46	4.83	4.47	4.24	4.07	3.95	3.85	3.78	3.72	3.62	3.52	3.42	3.37	3.31	3.26	3.20	3.14	3.08
11	6.72	5.26	4.63	4.28	4.04	3.88	3.76	3.66	3.59	3.53	3.43	3.33	3.23	3.17	3.12	3.06	3.00	2.94	2.88
12	6.55	5.10	4.47	4.12	3.89	3.73	3.61	3.51	3.44	3.37	3.28	3.18	3.07	3.02	2.96	2.91	2.85	2.79	2.72
13	6.41	4.97	4.35	4.00	3.77	3.60	3.48	3.39	3.31	3.25	3.15	3.05	2.95	2.89	2.84	2.78	2.72	2.66	2.60
14	6.30	4.86	4.24	3.89	3.66	3.50	3.38	3.29	3.21	3.15	3.05	2.95	2.84	2.79	2.73	2.67	2.61	2.55	2.49
15	6.20	4.77	4.15	3.80	3.58	3.41	3.29	3.20	3.12	3.06	2.96	2.86	2.76	2.70	2.64	2.59	2.52	2.46	2.40
16	6.12	4.69	4.08	3.73	3.50	3.34	3.22	3.12	3.05	2.99	2.89	2.79	2.68	2.63	2.57	2.51	2.45	2.38	2.32
17	6.04	4.62	4.01	3.66	3.44	3.28	3.16	3.06	2.98	2.92	2.82	2.72	2.62	2.56	2.50	2.44	2.38	2.32	2.25
18	5.98	4.56	3.95	3.61	3.38	3.22	3.10	3.01	2.93	2.87	2.77	2.67	2.56	2.50	2.44	2.38	2.32	2.26	2.19
19	5.92	4.51	3.90	3.56	3.33	3.17	3.05	2.96	2.88	2.82	2.72	2.62	2.51	2.45	2.39	2.33	2.27	2.20	2.13
20	5.87	4.46	3.86	3.51	3.29	3.13	3.01	2.91	2.84	2.77	2.68	2.57	2.46	2.41	2.35	2.29	2.22	2.16	2.09
21	5.83	4.42	3.82	3.48	3.25	3.09	2.97	2.87	2.80	2.73	2.64	2.53	2.42	2.37	2.31	2.25	2.18	2.11	2.04
22	5.79	4.38	3.78	3.44	3.22	3.05	2.93	2.84	2.76	2.70	2.60	2.50	2.39	2.33	2.27	2.21	2.14	2.08	2.00
23	5.75	4.35	3.75	3.41	3.18	3.02	2.90	2.81	2.73	2.67	2.57	2.47	2.36	2.30	2.24	2.18	2.11	2.04	1.97
24	5.72	4.32	3.72	3.38	3.15	2.99	2.87	2.78	2.70	2.64	2.54	2.44	2.33	2.27	2.21	2.15	2.08	2.01	1.94
25	5.69	4.29	3.69	3.35	3.13	2.97	2.85	2.75	2.68	2.61	2.51	2.41	2.30	2.24	2.18	2.12	2.05	1.98	1.91
26	5.66	4.27	3.67	3.33	3.10	2.94	2.82	2.73	2.65	2.59	2.49	2.39	2.28	2.22	2.16	2.09	2.03	1.95	1.88
27	5.63	4.24	3.65	3.31	3.08	2.92	2.80	2.71	2.63	2.57	2.47	2.36	2.25	2.19	2.13	2.07	2.00	1.93	1.85
28	5.61	4.22	3.63	3.29	3.06	2.90	2.78	2.69	2.61	2.55	2.45	2.34	2.23	2.17	2.11	2.05	1.98	1.91	1.83
29	5.59	4.20	3.61	3.27	3.04	2.88	2.76	2.67	2.59	2.53	2.43	2.32	2.21	2.15	2.09	2.03	1.96	1.89	1.81
30	5.57	4.18	3.59	3.25	3.03	2.87	2.75	2.65	2.57	2.51	2.41	2.31	2.20	2.14	2.07	2.01	1.94	1.87	1.79
40	5.42	4.05	3.46	3.13	2.90	2.74	2.62	2.53	2.45	2.39	2.29	2.18	2.07	2.01	1.94	1.88	1.80	1.72	1.64
60	5.29	3.93	3.34	3.01	2.79	2.63	2.51	2.41	2.33	2.27	2.17	2.06	1.94	1.88	1.82	1.74	1.67	1.58	1.48
120	5.15	3.80	3.23	2.89	2.67	2.52	2.39	2.30	2.22	2.16	2.05	1.94	1.82	1.76	1.69	1.61	1.53	1.43	1.31
∞	5.02	3.69	3.12	2.79	2.57	2.41	2.29	2.19	2.11	2.05	1.94	1.83	1.71	1.64	1.57	1.48	1.39	1.27	1.00

n_2 = degrees of freedom for denominator

27

Table 1.4 (Continued)

$F_{.99}(n_1, n_2)$

n_1 = degrees of freedom for numerator

n_2 = degrees of freedom for denominator

n_2 \ n_1	1	2	3	4	5	6	7	8	9	10	12	15	20	24	30	40	60	120	∞
1	4052	4999.5	5403	5625	5764	5859	5928	5982	6022	6056	6106	6157	6209	6235	6261	6287	6313	6339	6366
2	98.50	99.00	99.17	99.25	99.30	99.33	99.36	99.37	99.39	99.40	99.42	99.43	99.45	99.46	99.47	99.47	99.48	99.49	99.50
3	34.12	30.82	29.46	28.71	28.24	27.91	27.67	27.49	27.35	27.23	27.05	26.87	26.69	26.60	26.50	26.41	26.32	26.22	26.13
4	21.20	18.00	16.69	15.98	15.52	15.21	14.98	14.80	14.66	14.55	14.37	14.20	14.02	13.93	13.84	13.75	13.65	13.56	13.46
5	16.26	13.27	12.06	11.39	10.97	10.67	10.46	10.29	10.16	10.05	9.89	9.72	9.55	9.47	9.38	9.29	9.20	9.11	9.02
6	13.75	10.92	9.78	9.15	8.75	8.47	8.26	8.10	7.98	7.87	7.72	7.56	7.40	7.31	7.23	7.14	7.06	6.97	6.88
7	12.25	9.55	8.45	7.85	7.46	7.19	6.99	6.84	6.72	6.62	6.47	6.31	6.16	6.07	5.99	5.91	5.82	5.74	5.65
8	11.26	8.65	7.59	7.01	6.63	6.37	6.18	6.03	5.91	5.81	5.67	5.52	5.36	5.28	5.20	5.12	5.03	4.95	4.86
9	10.56	8.02	6.99	6.42	6.06	5.80	5.61	5.47	5.35	5.26	5.11	4.96	4.81	4.73	4.65	4.57	4.48	4.40	4.31
10	10.04	7.56	6.55	5.99	5.64	5.39	5.20	5.06	4.94	4.85	4.71	4.56	4.41	4.33	4.25	4.17	4.08	4.00	3.91
11	9.65	7.21	6.22	5.67	5.32	5.07	4.89	4.74	4.63	4.54	4.40	4.25	4.10	4.02	3.94	3.86	3.78	3.69	3.60
12	9.33	6.93	5.95	5.41	5.06	4.82	4.64	4.50	4.39	4.30	4.16	4.01	3.86	3.78	3.70	3.62	3.54	3.45	3.36
13	9.07	6.70	5.74	5.21	4.86	4.62	4.44	4.30	4.19	4.10	3.96	3.82	3.66	3.59	3.51	3.43	3.34	3.25	3.17
14	8.86	6.51	5.56	5.04	4.69	4.46	4.28	4.14	4.03	3.94	3.80	3.66	3.51	3.43	3.35	3.27	3.18	3.09	3.00
15	8.68	6.36	5.42	4.89	4.56	4.32	4.14	4.00	3.89	3.80	3.67	3.52	3.37	3.29	3.21	3.13	3.05	2.96	2.87
16	8.53	6.23	5.29	4.77	4.44	4.20	4.03	3.89	3.78	3.69	3.55	3.41	3.26	3.18	3.10	3.02	2.93	2.84	2.75
17	8.40	6.11	5.18	4.67	4.34	4.10	3.93	3.79	3.68	3.59	3.46	3.31	3.16	3.08	3.00	2.92	2.83	2.75	2.65
18	8.29	6.01	5.09	4.58	4.25	4.01	3.84	3.71	3.60	3.51	3.37	3.23	3.08	3.00	2.92	2.84	2.75	2.66	2.57
19	8.18	5.93	5.01	4.50	4.17	3.94	3.77	3.63	3.52	3.43	3.30	3.15	3.00	2.92	2.84	2.76	2.67	2.58	2.49
20	8.10	5.85	4.94	4.43	4.10	3.87	3.70	3.56	3.46	3.37	3.23	3.09	2.94	2.86	2.78	2.69	2.61	2.52	2.42
21	8.02	5.78	4.87	4.37	4.04	3.81	3.64	3.51	3.40	3.31	3.17	3.03	2.88	2.80	2.72	2.64	2.55	2.46	2.36
22	7.95	5.72	4.82	4.31	3.99	3.76	3.59	3.45	3.35	3.26	3.12	2.98	2.83	2.75	2.67	2.58	2.50	2.40	2.31
23	7.88	5.66	4.76	4.26	3.94	3.71	3.54	3.41	3.30	3.21	3.07	2.93	2.78	2.70	2.62	2.54	2.45	2.35	2.26
24	7.82	5.61	4.72	4.22	3.90	3.67	3.50	3.36	3.26	3.17	3.03	2.89	2.74	2.66	2.58	2.49	2.40	2.31	2.21
25	7.77	5.57	4.68	4.18	3.85	3.63	3.46	3.32	3.22	3.13	2.99	2.85	2.70	2.62	2.54	2.45	2.36	2.27	2.17
26	7.72	5.53	4.64	4.14	3.82	3.59	3.42	3.29	3.18	3.09	2.96	2.81	2.66	2.58	2.50	2.42	2.33	2.23	2.13
27	7.68	5.49	4.60	4.11	3.78	3.56	3.39	3.26	3.15	3.06	2.93	2.78	2.63	2.55	2.47	2.38	2.29	2.20	2.10
28	7.64	5.45	4.57	4.07	3.75	3.53	3.36	3.23	3.12	3.03	2.90	2.75	2.60	2.52	2.44	2.35	2.26	2.17	2.06
29	7.60	5.42	4.54	4.04	3.73	3.50	3.33	3.20	3.09	3.00	2.87	2.73	2.57	2.49	2.41	2.33	2.23	2.14	2.03
30	7.56	5.39	4.51	4.02	3.70	3.47	3.30	3.17	3.07	2.98	2.84	2.70	2.55	2.47	2.39	2.30	2.21	2.11	2.01
40	7.31	5.18	4.31	3.83	3.51	3.29	3.12	2.99	2.89	2.80	2.66	2.52	2.37	2.29	2.20	2.11	2.02	1.92	1.80
60	7.08	4.98	4.13	3.65	3.34	3.12	2.95	2.82	2.72	2.63	2.50	2.35	2.20	2.12	2.03	1.94	1.84	1.73	1.60
120	6.85	4.79	3.95	3.48	3.17	2.96	2.79	2.66	2.56	2.47	2.34	2.19	2.03	1.95	1.86	1.76	1.66	1.53	1.38
∞	6.63	4.61	3.78	3.32	3.02	2.80	2.64	2.51	2.41	2.32	2.18	2.04	1.88	1.79	1.70	1.59	1.47	1.32	1.00

"Reprinted from the United States Department of Commerce NBS Handbook 91.

F Distribution.   A final sampling pdf mentioned briefly here is the so-called F pdf. Given two independent sample variances s_1^2 and s_2^2 based on $\nu_1 = (n_1 - 1)$ and $\nu_2 = (n_2 - 1)$ degrees of freedom, respectively, the F-statistic

$$F = \frac{s_1^2/\sigma_1^2}{s_2^2/\sigma_2^2} \tag{1.6.10}$$

has a pdf of the form

$$f(F) = C_3 F^{(\nu_2-2)/2}\left[1 + \left(\frac{\nu_1 F}{\nu_2}\right)\right]^{-(\nu_1+\nu_2)/2} \tag{1.6.11}$$

where C_3 is a known function of ν_1 and ν_2. Since this is a two-parameter pdf, the cdf is more difficult to tabulate. Table 1.4 gives a tabulation for $P = 0.90$, 0.95, 0.975, and 0.99. The quantity tabulated is $F_{P(\nu_1, \nu_2)}$ defined by

$$\int_0^{F_P(\nu_1, \nu_2)} f(F)\, dF = P \tag{1.6.12}$$

EXAMPLE 1.M

To illustrate, if $\sigma_1^2 = \sigma_2^2$, $\nu_1 = 15$, and $\nu_2 = 4$, the probability is 0.975 that s_1^2/s_2^2 will be less than 8.66, or, equivalently, the probability is 0.025 that s_1^2/s_2^2 will exceed 8.66.   ■

1.7 STATISTICAL INFERENCE

Statistical inference is concerned with making statements about populations on the basis of samples drawn from the populations. There are two broad types of inference, both of which arise in later chapters. One is the problem of obtaining estimates of population parameters; the other concerns testing hypotheses about parameter values.

1.7.1 Point Estimation

For a given pdf containing a parameter θ, an *estimator* for θ is a specifically defined function of observations that yields a numerical value. The numerical value is called the *estimate* of θ. Any number of estimators might be used in a given instance, and the choice of which estimator to use depends on such factors as the complexity of applying a given estimator, knowledge about the pdf of the estimator, and its statistical properties.

In this book, four methods of estimation are used: method of moments, method of maximum likelihood, method of least squares, and method of generalized least squares. These estimation methods are treated briefly in the following sections, after which their statistical properties are discussed. Note that in a given instance, different estimation methods may provide the same estimator of a parameter.

Method of Moments.   Section 1.4.3 discusses the moments of a pdf. In an analogous fashion, a set of sample values $x_1, x_2, \ldots, x_n$ also has moments. The first moment about the origin is $\bar{x}$, defined by Eq. 1.6.1. The general kth moment about the mean is defined by

$$\tilde{\mu}_k = \frac{\sum\limits_{i=1}^{n} (x_i - \bar{x})^k}{n} \tag{1.7.1}$$

In the method of moments, the sample moments are set equal to the corresponding population moments, and the resulting equations are solved simultaneously to provide the estimators. For the case under discussion and with μ_k, $\sqrt{\beta_1}$, and β_2 defined by Eqs. 1.4.7 to 1.4.9, the moment estimators (ME's) of the parameters are

$$\tilde{\mu} = \bar{x} \tag{1.7.2}$$

$$\tilde{\mu}_2 = \tilde{\sigma}^2 = \frac{\sum\limits_{i=1}^{n} (x_i - \bar{x})^2}{n} \tag{1.7.3}$$

$$\sqrt{\tilde{\beta}_1} = \frac{\tilde{\mu}_3}{\tilde{\mu}_2^{1.5}} \tag{1.7.4}$$

$$\tilde{\beta}_2 = \frac{\tilde{\mu}_4}{\tilde{\mu}_2^2} \tag{1.7.5}$$

Other examples of estimators derived from the method of moments are encountered in later chapters.

Method of Maximum Likelihood.   For a given set of observations, $x_1, x_2, \ldots, x_n$, from some specified pdf, there is an associated likelihood of occurrence. If the form of the pdf is specified but a given parameter value for θ is not, then the maximum likelihood estimator (MLE) of θ is that value that produces the largest likelihood for the sample results.

This principle is most easily understood with a simple example. Given a sample of n independent observations, $x_1, x_2, \ldots, x_n$, from a normal pdf with mean μ and variance σ^2, the problem is to find the MLE's of μ and σ^2.

From Eq. 1.5.2, the pdf of any single observation, x_i, is

$$f(x_i) = \frac{1}{\sqrt{2\pi}\,\sigma} \exp\left[\frac{-(x_i - \mu)^2}{2\sigma^2}\right] \qquad (1.7.6)$$

By an extension of Eq. 1.3.4, for these n independent sample values, the pdf of the set of observations is

$$f(x_1, x_2, \ldots, x_n) = \frac{1}{(\sqrt{2\pi})^n \sigma^n} \exp\left[\frac{-\sum_{i=1}^{n}(x_i - \mu)^2}{2\sigma^2}\right] \qquad (1.7.7)$$

where $f(x_1, x_2, \ldots, x_n)$ is the likelihood of the sample values. The MLE's of μ and σ^2 are those values of the parameters that maximize this likelihood. It is simpler, and results are equivalent, to maximize the natural logarithm of the likelihood:

$$L = -0.5n \ln 2\pi - 0.5n \ln \sigma^2 - \frac{\sum_{i=1}^{n}(x_i - \mu)^2}{2\sigma^2} \qquad (1.7.8)$$

L is maximized by taking its partial derivatives with respect to μ and σ^2, setting the two expressions equal to zero, and solving the two equations simultaneously:

$$\frac{\partial L}{\partial \mu} = 0 \Rightarrow \sum_{i=1}^{n} x_i - n\mu = 0$$

$$\frac{\partial L}{\partial \sigma^2} = 0 \Rightarrow \frac{-0.5n}{\sigma^2} + \frac{\sum_{i=1}^{n}(x_i - \mu)^2}{2\sigma^4} = 0$$

From the first equation, the MLE of μ is

$$\hat{\mu} = \frac{\sum_{i=1}^{n} x_i}{n} = \bar{x} \qquad (1.7.9)$$

where the caret indicates that $\hat{\mu}$ is an estimator. The use of the caret is generally restricted to an MLE.

Inserting $\bar{x}$ in place of μ in the second equation and solving for σ^2 gives its MLE:

$$\hat{\sigma}^2 = \frac{\sum\limits_{i=1}^{n} (x_i - \bar{x})^2}{n} \tag{1.7.10}$$

or, equivalently,

$$\hat{\sigma}^2 = \frac{\sum\limits_{i=1}^{n} x_i^2 - \left(\sum\limits_{i=1}^{n} x_i\right)^2 \Big/ n}{n} = \frac{(n-1)s^2}{n} \tag{1.7.11}$$

where s^2 is defined by Eq. 1.6.2.

The method of maximum likelihood is also used frequently in later chapters.

Least Squares and Generalized Least Squares. Again given the series of sample values $x_1, x_2, \ldots, x_n$, suppose the expected value of x_i is a function of m unknown parameters, $g(\theta_1, \theta_2, \ldots, \theta_m)_i$. Then, the least squares estimates (LSE's) of the θ's are those values that minimize the sum of the squared differences between the x_i's and their expected values. Denoting this sum of squared differences by Q, the LSE's of the θ's are those values that minimize

$$Q = \sum_{i=1}^{n} \left[x_i - g(\theta_1, \theta_2, \ldots, \theta_m)_i\right]^2 \tag{1.7.12}$$

As an illustration, least squares estimation is commonly applied when a straight line is fit through a set of data. Formally, pairs of observations (x_i, y_i) are observed with

$$E(x_i) = a + by_i \tag{1.7.13}$$

The LSE's of the parameters a and b are those values that minimize

$$Q = \sum_{i=1}^{n} (x_i - a - by_i)^2 \tag{1.7.14}$$

It is a simple matter to show that the LSE's of a and b are

$$\hat{a} = \bar{x} - \hat{b}\bar{y}$$

$$\hat{b} = \frac{\sum\limits_{i=1}^{n} x_i y_i - \sum\limits_{i=1}^{n} x_i \sum\limits_{i=1}^{n} y_i / n}{\sum\limits_{i=1}^{n} y_i^2 - \left(\sum\limits_{i=1}^{n} y_i\right)^2 \Big/ n} \tag{1.7.15}$$

The LSE method assumes that all x_i's have the same sampling variance and that the covariance of x_i and x_j is zero for all pairs (i, j). When either assumption is not valid, the LSE method may be generalized to take into account the different sampling variances and the covariances between the pairs of observations. This leads to generalized least squares estimators (GLSE's) (3). The method of generalized least squares is presented in Section 7.4 and is then applied to a number of specific cases. It is not treated further in this section.

1.7.2 Properties of Estimators

Having considered four common methods of estimation, it is natural to ask which method is preferred in a given instance. This is not a problem when the contending methods provide the same estimators, but this does not always occur.

There is no "correct" method. One can, however, investigate properties of estimators and choose the one, or ones, with desirable properties. For simplicity, this discussion is restricted to a single parameter.

Unbiased Estimators. An estimator, $\hat{\theta}$, of a parameter, θ, is said to be unbiased if

$$E(\hat{\theta}) = \theta \tag{1.7.16}$$

It is desirable, but not essential, to have an unbiased estimator. If $E(\hat{\theta})$ differs from θ only slightly relative to the standard deviation of $\hat{\theta}$, the estimator may be quite satisfactory. Further, it may be possible to modify an estimator slightly to remove the bias without affecting its other desirable properties. As an example, as shown in Eq. 1.7.10, for the model considered the MLE of σ^2 is

$$\hat{\sigma}^2 = \frac{\sum\limits_{i=1}^{n} (x_i - \bar{x})^2}{n} \tag{1.7.17}$$

This is a biased estimate of σ^2 since it can be shown that

$$E(\hat{\sigma}^2) = \frac{(n-1)\sigma^2}{n} \tag{1.7.18}$$

However, the estimator can be made unbiased very simply, by using $(n-1)$ as the divisor rather than n; in other words, by using s^2 defined by Eq. 1.6.2 as the estimator.

Consistent Estimators. An estimator $\hat{\theta}$ of a parameter θ is said to be consistent if the difference $(\hat{\theta} - \theta)$ approaches zero with probability one as the sample size approaches infinity. A *consistent estimator* is unbiased in the limit. However, it does not necessarily follow that an unbiased estimator is a consistent estimator.

Efficient Estimators. In many estimation problems, it is possible to construct an estimator $\hat{\theta}$ of θ such that $\sqrt{n}(\hat{\theta} - \theta)$ has a normal pdf with zero mean as the sample size gets large. Of this class of estimators, those that have the smallest limiting variance are called *efficient estimators* of θ. By the previous definition of consistency, it is noted that efficient estimators are necessarily consistent.

Sufficient Estimators. Although a more formal definition of a sufficient estimator can be given, this property is most easily understood by noting that an estimator of a parameter is sufficient if it contains all the information in the sample about the parameter.

In closing this brief discussion on properties of estimators, it is noted that if the assumed model is valid, MLE's in general have good properties and cannot be essentially improved (4). Such estimators are not always used in practice, however, because they are sometimes difficult to develop. However, with the high computational capabilities now available, the user is now able to apply MLE's with relative ease where once it was impracticable to do so.

1.7.3   Interval Estimation

Although an estimate of a given parameter may be the best one available in some sense, it is not guaranteed to be close to the value of the parameter being estimated. A quantitative measure of the closeness of an estimate to the parameter is achieved by constructing a confidence interval on the parameter. A confidence interval is defined below followed by methods of constructing such intervals. First, exact methods for constructing confidence intervals are given, followed by approximate methods for large samples. Note that there are other types of statistical intervals in addition to confidence intervals; these other intervals are briefly described later in this section.

Confidence Interval. A confidence interval on a parameter θ is an interval constructed from the sample data, within which the true parameter value lies with predetermined probability or degree of confidence. The parameter θ is, of course, a constant and not a random variable. The end points of the interval are the random variables.

The confidence level or degree of confidence has the following interpretation. In any single application, a given interval either includes the parameter value or it does not. A 95% confidence level interval, say, on θ, is one that will, in the long run, be expected to include θ 95% of the time for all intervals constructed in the given manner.

A confidence interval may be one- or two-sided; that is, either a lower limit or an upper limit, or both, may be found. It is common practice for two-sided limits to be symmetrically placed about θ, in that the tail probabilities are equal, but this is not a requirement. In some instances, for example, in placing confidence limits on σ^2 when sampling from a normal population, the confidence interval has shorter length if equal probabilities are not assigned to the two tails. However, for simplicity in application, it is common to construct intervals with equal tail probabilities.

Exact Confidence Intervals. The method of constructing exact confidence intervals is as follows. If possible, a function of the sample statistic and the unknown parameter in question is found, this function having a known pdf that is independent of the parameter and of any other unknown parameter. A probability statement is then made about this function, and from this probability statement the confidence interval on the unknown parameter is derived. The procedure is best understood by an example.

EXAMPLE 1.N

In constructing a confidence interval on the mean, μ, of a normally distributed random variable based on a random sample of size n, the sample mean $\bar{x}$ and the sample variance s^2 are calculated from Eqs. 1.6.1 and 1.6.2, respectively. The required function of the test statistics and of the unknown parameter is the t statistic defined by Eq. 1.6.7. In a series of pilot runs on a denuder, the following caustic concentrations were observed:

$$
\begin{array}{cc}
68.5\% & 68.6\% \\
69.2\% & 68.8\% \\
68.2\% & 68.9\%
\end{array}
$$

Assuming the required assumptions to be valid, and calculating

$$\bar{x} = 68.700 \qquad s = 0.3464$$

the 95% confidence limits on μ are found by solving

$$\Pr\left(-2.571 < \frac{68.700 - \mu}{0.3464/\sqrt{6}} < 2.571\right) = 0.95$$

where 2.571 is read from Table 1.3 with 5 degrees of freedom and probability of 0.025 in each tail. Rewriting the probability equation to solve for μ, the 95% confidence interval on μ is

$$68.700 \pm \frac{(2.571)(0.3464)}{\sqrt{6}}$$

or $(68.336, 69.064)$. ■

EXAMPLE 1.O

These same data are used to construct a 90% confidence interval on σ^2 and hence also on σ. The statistic defined by Eq. 1.6.3 is the key statistic in this instance, and from Table 1.2, the probability equation is

$$\Pr\left[1.15 < \frac{(5)(0.1200)}{\sigma^2} < 11.07\right] = 0.90$$

Solving for σ^2 gives the interval. The lower limit is

$$\sigma^2 > \frac{(5)(0.1200)}{11.07} = 0.0542$$

Also,

$$\sigma^2 < \frac{(5)(0.1200)}{1.15} = 0.5217$$

Thus

$$(0.0542 < \sigma^2 < 0.5217) \Rightarrow (0.233 < \sigma < 0.722)$$ ■

It is not always possible to find a function needed to construct confidence intervals by this method. Confidence intervals can be constructed by a general method if the pdf of the estimator for the parameter is known. A further discussion of the general method is beyond the scope of this introductory material, and its application is not required in the later chapters. However, large-sample theory is used later. The construction of confidence intervals based on large-sample theory is presented next.

Confidence Intervals Based on Large-Sample Theory.   For large samples and under rather general conditions that are usually met in practice, the MLE $\hat{\theta}$ of a parameter θ with pdf $f(x; \theta)$ is approximately normally distributed with mean θ and with variance

$$\sigma_{\hat{\theta}}^2 = -\left\{nE\left[\frac{\partial^2 \ln f(x; \hat{\theta})}{\partial \theta^2}\right]\right\}^{-1} \tag{1.7.19}$$

where n is the sample size. Thus from Eq. 1.5.4, the approximate probability statement may be made, for a $100(1 - \alpha)\%$ confidence level,

$$\Pr\left(-z_{1-\alpha/2} < \frac{\hat{\theta} - \theta}{\sigma_{\hat{\theta}}} < z_{1-\alpha/2}\right) = 1 - \alpha \tag{1.7.20}$$

This probability statement may be solved for θ to yield the approximate 95% confidence interval on θ. The use of large-sample theory to construct confidence intervals in the multivariate case is discussed in Sections 5.4 and 6.4.

Other Statistical Intervals. Although the confidence interval is the only type of statistical interval to be constructed in later chapters, there are other kinds of statistical intervals. A *tolerance* interval is one that with a given level of confidence includes a specified proportion of the population (as compared to a parameter). Thus, tolerance interval statements involve two probability values, one referring to the level of confidence and the other to the proportion of the population that is covered. A *prediction* interval includes a future observation or observations, or specified functions of such observations with given probability.

1.7.4 Hypothesis Testing

In addition to finding point and interval estimates of parameters, an important part of statistical inference is the testing of hypotheses about population parameters. This subject is dealt with rather thoroughly in succeeding chapters for the general model being studied. In this section basic principles of hypothesis testing, exact tests, and approximate large-sample tests are discussed.

Principles of Hypothesis Testing. In its formal structure, hypothesis testing involves such terms as null and alternative hypotheses, type 1 and type 2 errors, α and β probability values, significance level, critical region, and test power. This section discusses these terms in narrative form; applications of hypothesis testing given in later chapters should help to further clarify the terms and their meanings.

A *null hypothesis*, often denoted by H_0, is established. A null hypothesis is a statement about the value of some parameter, or set of parameters, or more generally about pdf's from one or possibly several populations. For example, a null hypothesis might be that the random error variances associated with a given measurement process are equal to one another for several laboratories, or possibly, are equal to a given hypothesized value or set of laboratory-associated values.

To test the truth of the hypothesis, an experiment is run. A given statistic or set of statistics is calculated from the experimental data and used to reject H_0 or accept it. (It is preferable to speak of failing to reject a given hypothesis rather than accepting it, for it is difficult to establish the truth of a hypothesis. For example, a null hypothesis may be that a given coin is fair, i.e., has an equal probability of falling heads or tails. The experiment might be to toss the coin twice. Clearly, no matter what outcome occurs, there would be slight cause for rejection of the hypothesis, but neither would one have evidence to accept it. However, for simplicity in exposition, we speak of either rejecting or accepting the hypothesis in this further discussion.)

Based on the value of the test statistic(s), the null hypothesis will be rejected or accepted. If, in fact, H_0 is true and is accepted, or if it is false and is rejected, no decision error has been committed. There are two other possibilities that are decision errors:

1. H_0 is true, but it is rejected. This decision error is called a *type 1 error*.
2. H_0 is false but is accepted. This decision error is called a *type 2 error*.

One constructs the experiment to control the probabilities of committing these decision errors. The probability of committing a type 1 error is commonly denoted by α, called the *significance level* of the test. The probability of committing a type 2 error is usually denoted by β, which has meaning only if the *alternative hypothesis* is specified. An alternative hypothesis is a value outside the parameter region designated by H_0, usually denoted by H_1 or H_A. Because of the relationship between β and H_1, it is instructive to calculate β as a function of parameter values specified by the alternative hypothesis. The value of the complementary probability, $(1 - \beta)$, the probability of accepting H_1, is called the *test power* at H_1, and a plot of $(1 - \beta)$ versus H_1 is called a *power curve*. It is sometimes difficult to specify an alternative hypothesis. For example, H_0 may be that a random variable is normally distributed. Departure from normality may occur in many ways, and the most powerful test of H_0 will depend on the type of departure specified.

Specification of H_0, definition of the test statistics, and assignment of a value to β determine a *critical region*. If the test statistic falls within the critical region, H_0 is rejected.

Some exact tests are considered briefly in the next section.

Exact Tests of Hypotheses. Examples 1.N and 1.O, discussed earlier from the viewpoint of confidence interval estimation, are now considered from the viewpoint of hypothesis testing to illustrate two simple applications. As indicated earlier, later chapters include many additional applications.

EXAMPLE 1.P

Use the data of Example 1.N to test the null hypothesis:

$$H_0: \quad \mu \leq 68.0$$

against the alternative:

$$\mu > 68.0$$

From an intuitive viewpoint, H_0 will be rejected if $\bar{x}$, the estimate of μ, is too large. Formally, H_0 is rejected if $\bar{x} > c$ (c being the critical value, with $> c$ defining the critical region). The significance level of the test, α, is set equal, say, to 0.01. Thus we find c such that

$$\Pr(\bar{x} > c | \mu = 68.0) = 0.01$$

since the significance level is the probability of rejecting H_0 (occurs when $\bar{x} > c$), given that H_0 is true ($\mu = 68.0$, the upper boundary of the region within which H_0 is true). To find c, use is made of the fact that $(\bar{x} - \mu)\sqrt{n}/s$ is distributed as Students' t with $(n - 1)$ degrees of freedom. (See Eq. 1.6.7.) This fact was also utilized in finding the confidence interval, and there is a formal correspondence between the two kinds of statistical inference. Thus

$$\Pr(\bar{x} > c | \mu = 68.0) = \Pr\left[t > \left(\frac{c - 68.0}{s/\sqrt{n}} \right) \right] = 0.01$$

For $n = 6$, from Table 1.3,

$$\frac{c - 68.0}{s/\sqrt{6}} = 3.365$$

and hence $c = 68.0 + 1.3738s$.

Thus $\bar{x}$ and s are calculated from the sample data, and H_0 is rejected if $\bar{x} > c$. From the cited example,

$$\bar{x} = 68.700 \qquad s = 0.3464$$

Thus $c = 68.476$.

Since $\bar{x} > 68.476$, the null hypothesis H_0 is rejected. Note that this example is a one-sided test, that is, H_0 is rejected only if $\bar{x}$ is too large. One could also formulate the null hypothesis, H_0: $\mu = 68.0$, the alternative hypothesis region being $\mu \neq 68.0$. In this instance, H_0 would be rejected if $\bar{x}$ were either too small or too large. The mechanics of determining the two critical values c_1 and c_2 are analogous to those used to determine c, the probability equation now being

$$\Pr(\bar{x} < c_1 | \mu = 68.0) + \Pr(\bar{x} > c_2 | \mu = 68.0) = \alpha$$

where, unless otherwise specified, each indicated probability would be $\alpha/2$. ∎

EXAMPLE 1.Q

Following Example 1.O, test the null hypothesis: H_0: $\sigma^2 \leq 0.0625$ against the alternative: $\sigma^2 > 0.0625$. Setting $\alpha = 0.25$ in this example, the critical value c is found from the probability equation

$$\Pr(s^2 > c | \sigma^2 = 0.0625) = 0.025$$

To find c, recall from Eq. 1.6.3 that $u = (n - 1)s^2/\sigma^2$ has the chi-square pdf with $(n - 1)$ degrees of freedom. Thus the probability equation may be rewritten for this example:

$$\Pr\left(u > \frac{5c}{0.0625}\right) = 0.025$$

and from Table 1.2,

$$\frac{5c}{0.0625} = 12.83, \text{ or}$$

$$c = 0.1604$$

From the cited example,

$$s^2 = 0.1200$$

and, hence, H_0 is not rejected since $s^2 < c$. ∎

Other applications of exact tests of hypotheses are given in later chapters, in which it is also necessary, in some instances, to apply approximate tests based on large-sample theory because exact tests cannot always be formulated.

Large-Sample Tests.   Just as there are a number of estimation approaches for parameter estimators, there are various methods for constructing tests of hypotheses. The so-called likelihood ratio test, related to the estimation principle of maximum likelihood, usually leads to a very good test and is the basis for many of the common tests. Further, the likelihood ratio test is relevant to our development, for the large-sample tests utilized are based on the large-sample probability distribution of the likelihood ratio statistic.

Let $x_1, x_2, \ldots, x_n$ be an independently drawn random sample of size n from a population with pdf $f(x; \theta_1, \theta_2, \ldots, \theta_m)$. The null hypothesis H_0 is that the pdf belongs to a subspace ω of the entire parameter space Ω. For example, H_0 may be that $\theta_1 = \theta_2; \theta_3 = \theta_4 = \theta_5; \theta_6 = \cdots = \theta_m$ or it may be that $\theta_i = \theta_j$ for all i, j. Now, the likelihood of the sample is

$$L' = \prod_{i=1}^{n} f(x_i; \theta_1, \theta_2, \ldots, \theta_m) \tag{1.7.21}$$

Over the entire space Ω, the likelihood will at some point reach a maximum, corresponding to the MLE's of the θ's. This maximum is denoted by $L'(\hat{\Omega})$. Similarly, in the restricted subspace ω under H_0, L' will also have a maximum corresponding to the MLE's of the parameters in ω. This maximum is denoted by $L'(\hat{\omega})$. The likelihood ratio is

$$\frac{L'(\hat{\omega})}{L'(\hat{\Omega})}$$

which is a positive value with range 0 to 1. Again arguing intuitively, if this ratio is near 1, the data are supportive of the truth of H_0, while if the ratio differs appreciably from 1, rejection of H_0 is called for. The problem is to determine the critical value for the ratio corresponding to a fixed significance level α.

Under general conditions that are in fact satisfied in later chapters, and for large sample sizes, twice the natural logarithm of the likelihood ratio,

$$\lambda = -2[L(\hat{\omega}) - L(\hat{\Omega})] \tag{1.7.22}$$

is approximately distributed as chi-square with $(m - r)$ degrees of freedom, where r is the dimensionality of ω, that is, the number of parameters estimated under H_0. In Eq. 1.7.22, $L(\hat{\omega}) = \ln L'(\hat{\omega})$ and $L(\hat{\Omega}) = \ln L'(\hat{\Omega})$.

Thus if one knows the large sample pdf for λ, the critical value may be found for given α. Several illustrations of the large-sample likelihood ratio test are given in later chapters.

1.8 MATHEMATICAL MODELS

Statistical inference depends heavily on the underlying mathematical model, which includes a complete specification of the relationships among all the experimental variables, random and controlled; statements about the pdf's and their parameters; definitions of the populations; and any other information needed for valid inferential statistical analyses.

The importance of a careful model definition cannot be overstressed. Much confusion about conclusions reached in statistical analyses may be traced to poor or incomplete model definition. The method of analysis depends quite heavily on the assumed model, and violations of the assumptions of the model can lead to grossly incorrect conclusions.

Two competing motivations must be kept in mind when defining a model. On the one hand, the model should correspond closely to reality, a complete specification of which may be quite complex; on the other hand, the model should be simple enough to permit using existing methods of statistical inference to the extent possible, or modifications thereof that lead to workable solution methods. The problem is to construct a model that is an appropriate balance between these two extremes. Close cooperation between a researcher and data analyst may be required before settling on a model. In this age of high-speed computers, closer approximations of models to reality are achievable than was true in the past.

Because a body of statistical methods of inference is based on the assumption that the random variable is normally distributed, a common failing is to assume normality when such an assumption in fact is invalid. Since most of the inferential methodology in this book also assumes that the random variable has a normal pdf, one might suspect that I have chosen to compromise reality in order to work with a desirable model. However, this suspicion is unfounded since the emphasis in this book is on errors of measurement. A large body of experience exists in many fields of application that support the assertion that errors of measurement do, in fact, often have normal pdf's. It is also noted that some measurements may be made in a metric that does not provide normally distributed errors, but often in such cases data transformations can be made to achieve normality (see Example 4.D).

Having discussed briefly the importance of mathematical modeling, we now specify the general model that forms the basis for the methodology to be developed in this book.

1.8.1 Mathematical Model for Measurement Errors

The experimental situation that forms the basis for the remaining chapters was described in Section 1.1. For this situation, let

x_{ik} = observed measurement value for item k, method i, where

$k = 1, 2, \ldots, n; \qquad i = 1, 2, \ldots, N$

μ_k = true but unknown value for item k

The general model relating x_{ik} to μ_k is

$$x_{ik} = \alpha_i + \beta_i \mu_k + \varepsilon_{ik} \tag{1.8.1}$$

where α_i and β_i are parameters that jointly describe the measurement bias for method i, and where ε_{ik} is the random error of measurement committed in measuring item k with method i. It is assumed throughout that ε_{ik} is normally distributed with mean zero and variance σ_i^2. It is further assumed that

$$E(\varepsilon_{ik}) = 0 \qquad \text{for all } i, k \tag{1.8.2}$$

$$E(\varepsilon_{ik}^2) = \sigma_i^2 \qquad \text{for all } k \tag{1.8.3}$$

Further,

$$E(\varepsilon_{ik}\varepsilon_{jk}) = 0 \qquad \text{for all } k \text{ and for all pairs } (i, j) \tag{1.8.4}$$

and

$$E(\varepsilon_{ik}\varepsilon_{il}) = 0 \qquad \text{for all } i \text{ and for all pairs } (k, l) \tag{1.8.5}$$

With the base model given by Eq. 1.8.1, several variations on the model are considered in the chapters to follow. The following terminology is used in making these distinctions among the models.

- The models are said to have *common precisions* if σ_i^2 is the same for all i.
- If $\alpha_i = 0$ and $\beta_i = 1$, the method i is said to be *unbiased*.
- If $\alpha_i \neq 0$ but $\beta_i = 1$, method i is said to have a *constant bias*.
- If $\alpha_i \neq \alpha_j$ but $\beta_i = \beta_j = 1$, methods i and j are said to have a *constant relative bias*.
- If $\beta_i \neq 1$, method i is said to have a *nonconstant bias*.
- If $\beta_i \neq \beta_j$, methods i and j are said to have a *nonconstant relative bias*.

As long as the μ_k values are not known, one can never obtain estimates of the α's and β's, but only of their differences or ratios. For example, for a constant relative bias model one cannot distinguish between the case $\alpha_i = 5$, $\alpha_j = 2$ and the case $\alpha_i = 15$, $\alpha_j = 12$; one can only estimate the difference $(\alpha_i - \alpha_j)$. Similarly, only the ratio β_i/β_j can be estimated.

To continue with the terminology, it is important to specify the assumption about the μ_k's. A distinction is made between two extremes: a *fixed* and a *random* model. As the variations on the base model Eq. 1.8.1 are considered, it is seen that for some cases the distinction between a fixed and a random model is quite important, while for others it is not. In any event, the distinction must always be kept in mind.

- If μ_k is a constant for each k, that is, if it is not randomly selected from a population of values, then the model is called a *fixed* model.
- If μ_k is randomly selected from a population of values, μ_k being normally distributed with mean μ and variance σ_μ^2, then the model is called a *random* model.

Fixed and random models represent two extreme cases; there are an unlimited number of intermediate cases, since μ_k may be randomly selected from a variety of nonnormal populations. Clearly, all such cases cannot be treated. For those variations on the model for which the distinction between the fixed and random models is quite important, it would be advisable to assume the less restrictive fixed model unless the departures from normality can validly be assumed to be minor.

The model under study differs from an analysis-of-variance model, although readers familiar with the analysis of variance in its many forms might be inclined to use analysis of variance to analyze the data generated in this experimental situation. In an analysis of variance, σ^2 is assumed to be the same for all measurement methods, while for the model under study, this assumption is not made. Quite the contrary: the emphasis is on making inferences about the σ_i^2's and their relationships. Because analyses of variance cannot adequately be covered in a few pages, except for the simplest of models, and because an understanding of the analysis of variance is not needed in later chapters, no attempt is made to discuss this topic here. Many standard texts are available that describe analyses of variance quite thoroughly.

1.9 MATRIX ALGEBRA

In later chapters, use is sometimes made of matrix theory. A matrix is a rectangular array of quantities called the *elements* of the matrix. A matrix of dimension $n \times m$ is one with n rows and m columns. An $n \times m$ matrix may premultiply an $m \times p$ matrix to yield an $n \times p$ matrix as the product matrix. The element in row i, column j of the product matrix is found by multiplying row i of the $n \times m$ matrix into column j of the $m \times p$ matrix element by element and summing the products algebraically. For example,

$$\begin{pmatrix} 3 & 1 & -1 \\ 2 & -5 & 4 \end{pmatrix} \begin{pmatrix} 1 & -2 & 4 & 5 \\ 7 & -1 & 4 & -7 \\ -4 & 3 & -1 & 2 \end{pmatrix} = (?)$$

The row 1, column 1 element of the product matrix is

$$(3)(1) + (1)(7) + (-1)(-4) = 14$$

The row 2, column 3 element of the product matrix is

$$(2)(4) + (-5)(4) + (4)(-1) = -16$$

Continuing in this fashion, the product matrix has dimension 2×4 and is as follows:

$$\begin{pmatrix} 14 & -10 & 17 & 6 \\ -49 & 13 & -16 & 53 \end{pmatrix}$$

Matrix algebra is used in later chapters to solve systems of equations. To illustrate, consider the following system of three equations in three unknowns:

$$3x - 2y + 4z = 32$$

$$x + y - z = -6$$

$$-x - 4y + 2z = 20$$

In matrix notation, this system may be written as follows:

$$\begin{pmatrix} 3 & -2 & 4 \\ 1 & 1 & -1 \\ -1 & -4 & 2 \end{pmatrix} \begin{pmatrix} x \\ y \\ z \end{pmatrix} = \begin{pmatrix} 32 \\ -6 \\ 20 \end{pmatrix}$$

If one had a simple equation, say

$$3x = 5$$

and wanted to solve this for x, the solution would involve multiplying both sides of the equation by the inverse of 3, namely $\frac{1}{3}$, to give

$$\left(\tfrac{1}{3}\right)(3)(x) = \left(\tfrac{1}{3}\right)(5)$$

or

$$x = \tfrac{5}{3}$$

Matrix algebra proceeds in an analogous fashion. One finds the inverse of the matrix

$$\begin{pmatrix} 3 & -2 & 4 \\ 1 & 1 & -1 \\ -1 & -4 & 2 \end{pmatrix}$$

and premultiplies both sides of the matrix equation by this inverse to give the solution for

$$\begin{pmatrix} x \\ y \\ z \end{pmatrix}$$

The inverse of a matrix A is denoted by A^{-1} and is defined such that if one finds the product of matrices AA^{-1} or $A^{-1}A$, the resulting matrix is the so-called unit or identity matrix; the diagonal elements are all 1, and the off-diagonal elements are all 0. Only square matrices, for which the number of rows equals the number of columns, have inverses.

The problem is how to find the inverse matrix. It is not too difficult to find the inverse for 2×2 and 3×3 matrices, but larger dimensioned matrices call for computer assistance. Most libraries of computer programs contain sub-routines for inverting matrices of any reasonable size.

In continuing with the numerical example, the inverse of the 3×3 matrix is found to be

$$\tfrac{1}{16}\begin{pmatrix} 2 & 12 & 2 \\ 1 & -10 & -7 \\ 3 & -14 & -5 \end{pmatrix}$$

where the $\tfrac{1}{16}$ coefficient multiplies every element in the matrix.

To verify that this is the inverse of the earlier matrix, the reader may verify that the two matrices multiply together to give the unit matrix:

$$\tfrac{1}{16}\begin{pmatrix} 2 & 12 & 2 \\ 1 & -10 & -7 \\ 3 & -14 & -5 \end{pmatrix}\begin{pmatrix} 3 & -2 & 4 \\ 1 & 1 & -1 \\ -1 & -4 & 2 \end{pmatrix} = \begin{pmatrix} 1 & 0 & 0 \\ 0 & 1 & 0 \\ 0 & 0 & 1 \end{pmatrix}$$

Having found the inverse matrix, the solutions for x, y, and z are found by the following matrix calculation:

$$\begin{pmatrix} x \\ y \\ z \end{pmatrix} = \tfrac{1}{16} \begin{pmatrix} 2 & 12 & 2 \\ 1 & -10 & -7 \\ 3 & -14 & -5 \end{pmatrix} \begin{pmatrix} 32 \\ -6 \\ 20 \end{pmatrix}$$

$$= \begin{pmatrix} 2 \\ -3 \\ 5 \end{pmatrix}$$

Thus $x = 2$, $y = -3$, and $z = 5$ are the solutions, as can easily be verified.

REFERENCES

1. Feller, W. *An Introduction to Probability Theory and Its Applications*, Vol. I. New York: John Wiley and Sons; 1950; p. 45.

2. Hahn, G. J. and Shapiro, S. S. *Statistical Models in Engineering*. New York: John Wiley and Sons; 1967.

3. Aitken, A. C. "On Least Squares and Linear Combinations of Observations." *Proc. Royal Soc. Edinburgh*. **55**: 42–48; 1935.

4. Fisher, R. A. "Theory of Statistical Estimation." *Proc. Cambridge Phil. Soc.* **22**: 1925.

Chapter 2

THE COMMON PRECISION MODEL

2.1 INTRODUCTION

As defined in Section 1.8, for the common precision model, σ_i^2 is the same for all i. In Section 2.2, the common precision model is considered for $N = 2$ measurement methods on each of $k = 1, 2, \ldots, n$ items, in other words, for paired data. In Section 2.3, the case of $N \geq 3$ methods is discussed.

2.2 $N = 2$ MEASUREMENT METHODS

For the common precision model with $N = 2$, $\sigma_1^2 = \sigma_2^2 = \sigma_0^2$, and the problem is to make inferences about σ_0^2. The common precision model is often applicable when the same measurement method is used for both measurements. In developing the estimates of σ_0^2, the difference is calculated:

$$d_k = x_{1k} - x_{2k} \tag{2.2.1}$$

and the method of moments is used to estimate σ_0^2. The moment estimator (ME) is related to the maximum likelihood estimator (MLE) later.

For this model, Eq. 1.8.1 reduces to

$$x_{1k} = \alpha_1 + \mu_k + \varepsilon_{1k}$$

$$x_{2k} = \alpha_2 + \mu_k + \varepsilon_{2k} \tag{2.2.2}$$

where $E(\varepsilon_{1k}^2) = E(\varepsilon_{2k}^2) = \sigma_0^2$. Since μ_k does not appear once d_k is calculated, there is no distinction between the random and fixed models at this point in the discussion.

2.2.1 Point Estimate of σ_0^2

The difference d_k has the structure

$$d_k = (\alpha_1 - \alpha_2) + (\varepsilon_{1k} - \varepsilon_{2k}) \tag{2.2.3}$$

The mean of the d_k's, denoted by $\bar{d}$, has the structure

$$\bar{d} = (\alpha_1 - \alpha_2) + (\bar{\varepsilon}_1 - \bar{\varepsilon}_2) \tag{2.2.4}$$

where

$$\bar{d} = \frac{\sum\limits_{k=1}^{n} d_k}{n} \tag{2.2.5}$$

and

$$\bar{\varepsilon}_i = \frac{\sum\limits_{k=1}^{n} \varepsilon_{ik}}{n}; \qquad i = 1, 2 \tag{2.2.6}$$

Then the α's are eliminated by taking the difference

$$d_k - \bar{d} = (\varepsilon_{ik} - \varepsilon_{2k}) - (\bar{\varepsilon}_1 - \bar{\varepsilon}_2) \tag{2.2.7}$$

which has expected value

$$E(d_k - \bar{d})^2 = 2\sigma_0^2 + \frac{2\sigma_0^2}{n} - \frac{4\sigma_0^2}{n}$$

$$= \frac{2\sigma_0^2(n-1)}{n} \tag{2.2.8}$$

where the $4\sigma_0^2/n$ term is the sum of the covariances between ε_{1k} and $\bar{\varepsilon}_1$ and ε_{2k} and $\bar{\varepsilon}_2$. It follows immediately from Eq. 2.2.8 that the moment estimate of σ_0^2 is the mean of $(d_k - \bar{d})^2$ divided by $2(n-1)/n$:

$$\tilde{\sigma}_0^2 = \frac{\sum\limits_{k=1}^{n} (d_k - \bar{d})^2}{2(n-1)} = \frac{s_d^2}{2} \tag{2.2.9}$$

[Note: In calculating $\sum_{k=1}^{n}(d_k - \bar{d})^2$, the equivalent form,

$$\sum_{k=1}^{n} d_k^2 - \frac{\left(\sum\limits_{k=1}^{n} d_k\right)^2}{n}$$

may be used.]

From Eq. 2.2.3, it is seen that d_k is a normally distributed random variable with mean $(\alpha_1 - \alpha_2)$ and variance $2\sigma_0^2$. It is well known that the MLE of $2\sigma_0^2$ is

$$2\hat{\sigma}_0^2 = \frac{\sum\limits_{k=1}^{n}(d_k - \bar{d})^2}{n} \tag{2.2.10}$$

and, hence, the MLE of $\hat{\sigma}_0^2$ is the right-hand side (RHS) of Eq. 2.2.10 divided by 2. Note that this is $(n - 1)/n$ times the ME given in Eq. 2.2.9. The ME is unbiased.

It is noted that

$$s_d^2 = s_1^2 + s_2^2 - 2s_{12} \tag{2.2.11}$$

where s_1^2 is defined by Eq. 1.6.2 and where

$$s_{12} = \frac{\sum x_{1k}x_{2k} - \sum x_{1k}\sum x_{2k}/n}{n - 1} \tag{2.2.12}$$

the summations running from 1 to n. From Eq. 2.2.2, it follows that for a random model,

$$E(s_i^2) = \sigma_\mu^2 + \sigma_0^2; \quad i = 1, 2 \tag{2.2.13}$$

$$E(s_{12}) = \sigma_\mu^2 \tag{2.2.14}$$

and hence, the moment estimate of σ_0^2 is

$$\tilde{\sigma}_0^2 = \frac{s_1^2 + s_2^2 - 2s_{12}}{2} \tag{2.2.15}$$

the same as Eq. 2.2.9. The advantage of this formulation is that Eq. 2.2.14 provides an estimator for σ_μ^2 for a random model, a parameter that may be of interest in some cases (see Example 2.B).

It is also noted that if $\alpha_1 = \alpha_2$, then the moment estimate of σ_0^2 is simply

$$\tilde{\sigma}_0^2 = \frac{\sum\limits_{k=1}^{n} d_k^2}{2n} \tag{2.2.16}$$

Equations 2.2.9 and 2.2.16 both provide unbiased estimates if $\alpha_1 = \alpha_2$. However, if $\alpha_1 \neq \alpha_2$, then the estimate given by Eq. 2.2.16 is clearly biased.

Unless one is quite certain that $\alpha_1 = \alpha_2$, it is safer to use the estimate produced by Eq. 2.2.9; all that one gains in using Eq. 2.2.16 is one degree of freedom, an unimportant advantage unless n is small. In what follows, the estimator from Eq. 2.2.9 is used; the modifications that apply for the estimator from Eq. 2.2.16 are obvious.

2.2.2 Confidence Interval Estimate of σ_0^2

To obtain a confidence interval estimate of σ_0^2, use is made of the fact that $(n-1)s_d^2/2\sigma_0^2$ is distributed as chi-square with $(n-1)$ degrees of freedom. Therefore, the following probability statement may be made:

$$\Pr\left(\chi_{p_1}^2(n-1) < \frac{(n-1)s_d^2}{2\sigma_0^2} < \chi_{p_2}^2(n-1)\right) = p_2 - p_1 \quad (2.2.17)$$

where $\chi_{p_i}^2(\nu)$ is defined by

$$\int_0^{\chi_{p_i}^2(\nu)} \chi^2(\nu)\, d\chi^2 = p_i \quad (2.2.18)$$

and where $\chi^2(\nu)$ denotes the chi-square probability density function (pdf) with ν degrees of freedom. Simply stated, the area under the chi-square ν degrees of freedom curve from 0 to $\chi_{p_i}^2(\nu)$ is p_i.

From Eq. 2.2.17, the $100(p_2 - p_1)\%$ confidence limits on σ_0^2 are

$$\frac{(n-1)s_d^2}{2\chi_{p_2}^2(n-1)} < \sigma_0^2 < \frac{(n-1)s_d^2}{2\chi_{p_1}^2(n-1)} \quad (2.2.19)$$

Note that if, for example, 95% confidence limits are to be constructed, many combinations of p_1 and p_2 correspond to $(p_2 - p_1) = 0.95$. It is conventional to set $p_1 = 1 - p_2$, that is, to place equal probabilities in the two tails.

2.2.3 Testing Hypotheses About σ_0^2

To test the null hypothesis,

$$H_0: \quad \sigma_0^2 = \sigma_H^2$$

against the alternative hypothesis,

$$H_A: \quad \sigma_0^2 = \sigma_A^2$$

Table 2.1 Duplicate Analyses of Percentage Uranium for 18 Fuel Pellets (Coded)[a]

Pellet Number	x_{1k}	x_{2k}	d_k	Pellet Number	x_{1k}	x_{2k}	d_k
1	64	67	−3	10	71	68	3
2	48	69	−21	11	67	46	21
3	60	78	−18	12	82	37	45
4	105	71	34	13	82	42	40
5	97	72	25	14	51	58	−7
6	58	59	−1	15	62	59	3
7	64	50	14	16	50	52	−2
8	60	65	−5	17	49	31	18
9	69	58	11	18	43	55	−12

[a] $n = 18$ and $s_d^2 = 372.64 \times 10^{-6}$.

one again makes use of the known distribution of s_d^2. Specifically, suppose the test is one-sided, that is, $\sigma_A^2 > \sigma_H^2$. Then, for significance level p_1, the null hypothesis is rejected if

$$\frac{(n-1)s_d^2}{2\sigma_H^2} > \chi_{1-p_1}^2(n-1) \qquad (2.2.20)$$

or, equivalently, if

$$s_d^2 > \frac{2\sigma_H^2 \chi_{1-p_1}^2(n-1)}{n-1} \qquad (2.2.21)$$

EXAMPLE 2.A

Sintered UO_2 fuel pellets for use in a nuclear reactor are randomly selected from a production stream. The pellets are split in half, with one half analyzed immediately for percentage uranium and the other half retained for later analysis by the same analytical technique. The data for 18 pellets are listed in Table 2.1. These data are in percentage of uranium, and are coded for simplification by subtracting 88% from each value and then multiplying by 1000. Thus the first entry for x_{1k}, 64, corresponds to a percentage uranium value of 88.064%.

From Eq. 2.2.9, the estimate of σ_0^2 is

$$\tilde{\sigma}_0^2 = 186.32 \times 10^{-6}$$

and the estimate of σ_0 is 0.0136% uranium.

A 95% confidence interval is constructed on σ_0 by applying Eq. 2.2.19 to find the limits on σ_0^2 and then extracting the square roots of these limits. For $p_1 = 0.025$ and $p_2 = 0.975$, from Table 1.2 with $(n - 1) = 17$ degrees of freedom, we find

$$\chi^2_{0.025}(17) = 7.56$$

$$\chi^2_{0.975}(17) = 30.19$$

From Eq. 2.2.19, the 95% confidence limits on σ_0^2 are

$$\frac{6335 \times 10^{-6}}{(2)(30.19)} < \sigma_0^2 < \frac{6335 \times 10^{-6}}{(2)(7.56)}$$

$$1.049 \times 10^{-4} < \sigma_0^2 < 4.190 \times 10^{-4}$$

and the corresponding limits on σ_0 are

$$0.0102\% \ U < \sigma_0 < 0.0205\% \ U$$

That is, with 95% confidence, the true value of the parameter σ_0 lies between 0.0102% U and 0.0205% U.

Finally, suppose that the analyst claims that the standard deviation for this particular analytical method is 0.01% U, and we want to test the null hypothesis that $\sigma_0 = 0.01$ against the alternative hypothesis that $\sigma_0 > 0.01$. Let the significance level p_1 be 0.05. From Eq. 2.2.21, with

$$\chi^2_{0.95}(17) = 27.59 \qquad \text{and} \qquad \sigma_H = 0.01$$

the hypothesis is rejected if

$$s_d^2 > \frac{(2)(0.0001)(27.59)}{17} \qquad \text{or} \qquad 325 \times 10^{-6}$$

Since $s_d^2 = 373 \times 10^{-6}$, greater than the critical value, the hypothesis is rejected; the data do not support the analyst's contention that the standard deviation is 0.01% U. ∎

Example 2.B

Bennett and Franklin (1) present data representing duplicate results obtained by 38 students on a standard sample containing 29.27% iron. The data are reproduced in Table 2.2 with 25 subtracted from each number for convenience.

For this example, in the model Eq. 2.2.2, μ_k is the mean result for student k, and $(\alpha_1 - \alpha_2)$ is the relative bias between results 1 and 2. This relative bias

Table 2.2   Percent Iron − 25

Student	x_{1k}	x_{2k}	Difference d_k	Student	x_{1k}	x_{2k}	Difference d_k
1	4.41	4.32	0.09	20	4.27	4.25	0.02
2	4.27	4.39	−0.12	21	4.09	3.98	0.11
3	3.87	3.87	0.00	22	3.92	3.99	−0.07
4	4.04	4.03	0.01	23	4.26	4.15	0.11
5	4.73	4.96	−0.23	24	4.00	3.90	0.10
6	3.79	3.89	−0.10	25	4.40	4.23	0.17
7	4.30	4.20	0.10	26	3.96	4.04	−0.08
8	4.13	4.13	0.00	27	3.95	4.10	−0.15
9	4.21	4.05	0.16	28	3.90	3.97	−0.07
10	4.09	4.22	−0.13	29	4.19	4.20	−0.01
11	3.76	3.76	0.00	30	4.09	4.23	−0.14
12	4.33	4.24	0.09	31	4.07	4.10	−0.03
13	4.66	4.58	0.08	32	4.24	4.13	0.11
14	4.49	4.60	−0.11	33	4.27	4.20	0.07
15	4.42	4.39	0.03	34	4.85	4.80	0.05
16	4.32	4.25	0.07	35	4.35	4.45	−0.10
17	4.21	4.25	−0.04	36	4.32	4.11	0.21
18	4.30	4.08	0.22	37	4.20	4.42	−0.22
19	4.67	4.62	0.05	38	4.40	4.36	0.04

may logically be assumed to be zero, but since n is so large, to be on the safe side this assumption will not be made. It is further assumed that all students have the same precision σ_0^2.

For these data,

$$s_d^2 = 0.012624$$

and from Eq. 2.2.9,

$$\tilde{\sigma}_0^2 = 0.006312; \qquad \tilde{\sigma}_0 = 0.079\%$$

A 99% confidence interval is constructed on σ_0^2 by applying Eq. 2.2.19. For $p_1 = 0.005$ and $p_2 = 0.995$, from Table 1.2 with 37 degrees of freedom,

$$\chi_{0.005}^2(37) = 18.59 \qquad \chi_{0.995}^2(37) = 62.88$$

From Eq. 2.2.19, the 99% confidence limits on σ_0^2 are

$$\frac{0.467088}{(2)(62.88)} < \sigma_0^2 < \frac{0.467088}{(2)(18.59)}$$

$$0.033714 < \sigma_0^2 < 0.012563$$

and the corresponding limits on σ_0 are

$$0.0609 < \sigma_0 < 0.1121$$

Assuming that the 38 students are randomly and independently selected from a large population of students of like training and experience, and that μ_k is normally distributed with mean μ and variance σ_μ^2 (i.e., the model is random), then by Eq. 2.2.14, σ_μ^2 is estimated by s_{12}. From Table 2.2 and Eq. 2.2.12,

$$\tilde{\sigma}_\mu^2 = s_{12} = 0.057879$$

$$\tilde{\sigma}_\mu = 0.241\%$$

and note that the difference among students, described by σ_μ, is much larger than the difference between duplicate results, described by σ_0.   ∎

2.2.4   Estimation of Relative Biases

For the model Eq. 2.2.2, it is not possible to obtain estimates of α_1 and α_2 unless the values of the μ_k are known. It follows from Eq. 2.2.4 that the difference $(\alpha_1 - \alpha_2)$ is estimated by $\bar{d}$, since $\bar{\varepsilon}_1$ and $\bar{\varepsilon}_2$ both have zero expected values.

If, in a given application, the measurement methods are assumed to be randomly selected from an infinite population of methods (for example, they may represent two analysts performing the measurement, with the inference applicable to the population of analysts), then α_i is the realization of a random variable with mean and variance:

$$E(\alpha_i) = 0; \qquad i = 1, 2 \tag{2.2.22}$$

$$E(\alpha_i^2) = \sigma_\alpha^2 \tag{2.2.23}$$

and one wants to estimate σ_α^2. Again, the moment estimate of σ_α^2 follows immediately from Eq. 2.2.4 since

$$E(\bar{d}^2) = 2\sigma_\alpha^2 + \frac{2\sigma_0^2}{n} \tag{2.2.24}$$

With σ_0^2 estimated by Eq. 2.2.9, it is replaced by this estimate, $\tilde{\sigma}_0^2$, and hence the estimate of σ_α^2 is

$$\tilde{\sigma}_\alpha^2 = 0.5\bar{d}^2 - \frac{\tilde{\sigma}_0^2}{n} \tag{2.2.25}$$

It is noted that $\tilde{\sigma}_\alpha^2$ may be negative, in which case σ_α^2 would be assigned the value 0.

2.3 $N \geq 3$ MEASUREMENT METHODS

The discussion for $N \geq 3$ assumes that $\alpha_1 \neq \alpha_2 \neq \cdots \neq \alpha_N$, but the results are also applicable when all α's are equal.

2.3.1 Point Estimate of σ_0^2

Two approaches to estimating σ_0^2 are presented that are equivalent, in that they will always yield the same estimates of σ_0^2 and the other parameters.

Approach 1: Paired Differences.   Paired differences is an extension of the approach used for $N = 2$; it is presented here because of its relationship to an approach used in Chapter 3. The model is

$$x_{ik} = \alpha_i + \mu_k + \varepsilon_{ik}; \qquad i = 1, 2, \ldots, N$$

$$k = 1, 2, \ldots, n \tag{2.3.1}$$

The $N(N-1)/2$ columns of differences are formed:

$$d_{ijk} = x_{ik} - x_{jk} \tag{2.3.2}$$

Then, applying Eq. 2.2.9, an estimate of σ_0^2 is found for each pair of methods. For methods i and j,

$$\tilde{\sigma}_{0,ij}^2 = \frac{s_{dij}^2}{2(n-1)} \tag{2.3.3}$$

Table 2.3   Analysis of Variance Table for Two-Way Crossed-Classification Model

Source of Variation	Degrees of Freedom	Mean Square	Expected Mean Square
Measurement methods	$(N-1)$	M_1	$\sigma_0^2 + n\sigma_\alpha^2$
Items measured	$(n-1)$	M_2	$\sigma_0^2 + N\sigma_\mu^2$
Interaction = random error	$(N-1)(n-1)$	M_0	σ_0^2

The estimate of σ_0^2 is then found by averaging the $\tilde{\sigma}_{0,ij}^2$ values over the $N(N-1)/2$ pairs of methods:

$$\tilde{\sigma}_0^2 = \frac{2 \sum\limits_{i=1}^{N-1} \sum\limits_{j>i}^{N} \tilde{\sigma}_{0,ij}^2}{N(N-1)} \tag{2.3.4}$$

Approach 2: Analysis of Variance.   The data structure is that of a two-way crossed-classification analysis of variance with one observation per cell (2). The schematic analysis of variance table is given in Table 2.3.

The estimate of σ_0^2, namely the mean square M_0, is identical to the estimate from Eq. 2.3.4. Thus if one is interested in estimating σ_0^2, one may use either the paired differences or the analysis of variance approach to obtain identical estimators of the parameters.

2.3.2   Confidence Interval Estimate of σ_0^2

Confidence limits on σ_0^2 are most easily derived from the analysis of variance. The quantity $(N-1)(n-1)M_0/\sigma_0^2$ is distributed as chi-square with $(N-1)(n-1)$ degrees of freedom. Thus with $\chi_{p_i}^2[(N-1)(n-1)]$ defined by Eq. 2.2.18, the inequalities that lead to the confidence limits on σ_0^2 are

$$\chi_{p_1}^2[(N-1)(n-1)] < \frac{(N-1)(n-1)M_0}{\sigma_0^2} < \chi_{p_2}^2[(N-1)(n-1)]$$

$$\tag{2.3.5}$$

The $100(p_2 - p_1)\%$ confidence limits on σ_0^2 are therefore

$$\frac{(N-1)(n-1)M_0}{\chi_{p_2}^2[(N-1)(n-1)]} < \sigma_0^2 < \frac{(N-1)(n-1)M_0}{\chi_{p_1}^2[(N-1)(n-1)]} \tag{2.3.6}$$

2.3.3   Testing Hypotheses About σ_0^2

Following the argument of Section 2.3.2, to test the null hypothesis,

$$H_0: \quad \sigma_0^2 = \sigma_H^2$$

against the alternative hypothesis,

$$H_A: \quad \sigma_0^2 = \sigma_A^2, \quad \text{with} \quad \sigma_A^2 > \sigma_H^2$$

we see that H_0 will be rejected at a significance level p_1 if

$$\frac{(N-1)(n-1)M_0}{\sigma_H^2} > \chi_{1-p_1}^2[(N-1)(n-1)] \qquad (2.3.7)$$

or if

$$M_0 > \frac{\sigma_H^2 \chi_{1-p_1}^2[(N-1)(n-1)]}{(N-1)(n-1)} \qquad (2.3.8)$$

2.3.4   Estimation of Relative Biases

In estimating the relative biases for Approach 1, the paired difference approach, it follows from Eqs. 2.3.1 and 2.3.2 that

$$\ddot{\alpha}_i - \tilde{\alpha}_j = \bar{d}_{ij}$$

As was true for $N = 2$, it is not possible to obtain estimates of each α_i. However, if the arbitrary constraint that $\sum_{k=1}^{N} \alpha_k = 0$ is imposed, then all the α_k's may be estimated. These are interpreted as relative biases. The moment estimates are found to be

$$\tilde{\alpha}_i = \frac{\displaystyle\sum_{\substack{j \neq i}}^{N} \bar{d}_{ij}}{N} \qquad (2.3.9)$$

where

$$\bar{d}_{ij} = -\bar{d}_{ji} \qquad (2.3.10)$$

If the population of methods is assumed to be infinite, such that Eqs. 2.2.22 and 2.2.23 hold true, then one estimate of σ_α^2 may be found for each pair of methods, following Eq. 2.2.25.

$$\tilde{\sigma}_{\alpha_{ij}}^2 = 0.5\bar{d}_{ij}^2 - \frac{\tilde{\sigma}_0^2}{n} \qquad (2.3.11)$$

The estimate of σ_α^2 is then found by averaging the $\tilde{\sigma}_{\alpha,ij}^2$ values over all $N(N-1)/2$ pairs of methods:

$$\tilde{\sigma}_\alpha^2 = \frac{\displaystyle\sum_{i=1}^{N-1}\sum_{j>i}^{N} \bar{d}_{ij}^2}{N(N-1)} - \frac{\tilde{\sigma}_0^2}{n} \qquad (2.3.12)$$

In relating this estimate of σ_α^2 to the estimates of the α's in Eq. 2.3.9, it is noted that for finite populations the usual definition of the variance σ_α^2 is

$$\sigma_\alpha^2 = \frac{\displaystyle\sum_{k=1}^{N} \alpha_k^2}{N-1} \qquad (2.3.13)$$

If the individual α's given by Eq. 2.3.9 are squared and summed, the sum will be precisely that given by Eq. 2.3.12, except for the $\tilde{\sigma}_0^2/n$ term, as illustrated in the example to follow.

This apparent discrepancy requires some explanation. Note that Eq. 2.3.13 is the definition of σ_α^2 for the finite model. It does not follow that the estimate of this quantity is found by simply replacing the unknown α_k's by their estimates given by Eq. 2.3.9, since this procedure yields a biased estimate of σ_α^2. Rather, noting that

$$E\left(\frac{\displaystyle\sum_{k=1}^{N} \tilde{\alpha}_k^2}{N-1}\right) = \sigma_\alpha^2 + \frac{\sigma_0^2}{n} \qquad (2.3.14)$$

it follows that the unbiased estimate of σ_α^2 for the finite model is

$$\tilde{\sigma}_\alpha^2 = \frac{\displaystyle\sum_{k=1}^{N} \tilde{\alpha}_k^2}{N-1} - \frac{\tilde{\sigma}_0^2}{n} \qquad (2.3.15)$$

This is identical to the estimate given by Eq. 2.3.12 for the infinite model, as illustrated in the example to follow. Thus whether the model is finite or infinite, the estimate of σ_α^2 is the same, but the interpretation of the parameter is not.

Table 2.4   Recorded Weights, in Kilograms, of Eight Containers
of UO_2 Weighed on Four Scales

Container	Scale 1	Scale 2	Scale 3	Scale 4
1	20.146	20.121	20.119	20.099
2	22.096	22.051	22.086	22.061
3	21.144	21.094	21.107	21.094
4	21.488	21.431	21.470	21.439
5	21.922	21.882	21.900	21.857
6	18.703	18.672	18.691	18.675
7	20.447	20.413	20.413	20.389
8	24.967	24.920	24.949	24.944

For the analysis of variance approach of Section 2.3.1, the estimate of σ_α^2 derived from Table 2.3 is

$$\tilde{\sigma}_\alpha^2 = \frac{M_1 - M_0}{n} \qquad (2.3.16)$$

The estimator is identical to the estimator from Eq. 2.3.12 or that from Eq. 2.3.15 based on the paired difference approach. Thus as was true for the estimation of σ_0^2, it makes no difference whether the paired difference or the analysis of variance approach is used.

EXAMPLE 2.C

In a nuclear fuel fabrication facility, eight containers of uranium dioxide powder were weighed on each of four scales. The four scales are assumed to have the same measurement precision, that is, $\sigma_i^2 = \sigma_0^2$ for all i. The data are given in Table 2.4. The paired difference approach described in Approach 1 of Section 2.3.1 is used first. The d_{ijk} values of Eq. 2.3.2 are calculated. These are given in Table 2.5, expressed in grams UO_2.

In applying Eq. 2.3.3 for all pairs of scales, the six individual estimates of σ_0^2, in grams^2 of UO_2, are as follows:

$$\tilde{\sigma}_{0,12}^2 = 56.7768 \qquad \tilde{\sigma}_{0,23}^2 = 113.9196$$

$$\tilde{\sigma}_{0,13}^2 = 47.8214 \qquad \tilde{\sigma}_{0,24}^2 = 167.8214$$

$$\tilde{\sigma}_{0,14}^2 = 105.9911 \qquad \tilde{\sigma}_{0,34}^2 = 67.4911$$

The overall estimate is then given by Eq. 2.3.4 as the average of these six estimates.

$$\tilde{\sigma}_0^2 = 93.3036 \text{ grams}^2 \rightarrow \tilde{\sigma}_0 = 9.66 \text{ grams}$$

Table 2.5 Calculated d_{ijk} Values for Example 2.C

k	d_{12k}	d_{13k}	d_{14k}	d_{23k}	d_{24k}	d_{34k}
1	25	27	47	2	22	20
2	45	10	35	-35	-10	25
3	50	37	50	-13	0	13
4	57	18	49	-39	-8	31
5	40	22	65	-18	25	43
6	31	12	28	-19	-3	16
7	34	34	58	0	24	24
8	47	18	23	-29	-24	5
$\bar{d}_{ij}$ = mean =	41.125	22.250	44.375	-18.875	3.250	22.125

The six estimates are not all mutually independent here or in later applications, but such lack of independence is taken into account in making inferences.

Turning attention now to the estimates of the relative biases for a finite model, application of Eq. 2.3.9 gives the following results, in grams:

$$\tilde{\alpha}_1 - \tilde{\alpha}_2 = 41.125 \qquad \tilde{\alpha}_2 - \tilde{\alpha}_3 = -18.875$$

$$\tilde{\alpha}_1 - \tilde{\alpha}_3 = 22.250 \qquad \tilde{\alpha}_2 - \tilde{\alpha}_4 = 3.250$$

$$\tilde{\alpha}_1 - \tilde{\alpha}_4 = 44.375 \qquad \tilde{\alpha}_3 - \tilde{\alpha}_4 = 22.125$$

If the constraint is imposed that the α_k sum to zero over these four scales, then Eq. 2.3.10 is applied to give

$$\tilde{\alpha}_1 = \frac{41.125 + 22.250 + 44.375}{4} = 26.9375 \text{ grams}$$

$$\tilde{\alpha}_2 = \frac{-41.125 - 18.875 + 3.250}{4} = -14.1875 \text{ grams}$$

$$\tilde{\alpha}_3 = 4.6875 \text{ grams}$$

$$\tilde{\alpha}_4 = -17.4375 \text{ grams}$$

Note that these sum to zero, and that σ_α^2 given by Eq. 2.3.13, the definition of a population variance when the population is finite, is estimated to be 405.9881 grams^2 from Eq. 2.3.15.

Table 2.6   Analysis of Variance Table for Example 2.C

Source of Variation	Degrees of Freedom	Mean Square (grams² UO₂)
Measurement methods	3	$M_1 = 3341.23$
Items measured	7	$M_2 = 13,309,264$
Random error	21	$M_0 = 93.30$

Assume now that the model is infinite, and the problem is to estimate σ_α^2 rather than α_k for all k. In applying Eq. 2.3.12, the estimate of σ_α^2 is

$$\tilde{\sigma}_\alpha^2 = \frac{(41.125)^2 + \cdots + (22.125)^2}{12} - \frac{93.3036}{8}$$

$$= 417.6510 - 11.6630 = 405.9881 \text{ grams}^2$$

$$\tilde{\sigma}_\alpha = 20.149 \text{ grams}$$

Comparing this result with that for the finite model shows that they are identical.

The data of Table 2.4 are now analyzed by the analysis of variance described in Approach 2 of Section 2.3.1. The analysis of variance table, given schematically as Table 2.3, is shown in Table 2.6 for this example. Note that M_0 as the estimator of σ_0^2 is identical to the estimator given by Eq. 2.3.4 in the paired difference approach. Further, from Eq. 2.3.16,

$$\tilde{\sigma}_\alpha^2 = \frac{3341.23 - 93.30}{3} = 405.99 \text{ grams}^2$$

identical to the values given by Eqs. 2.3.12 or 2.3.15.

To obtain the confidence interval estimate of σ_0^2, apply Eq. 2.3.6. In finding 95% confidence limits, p_2 is 0.975 and p_1 is 0.025. There are $(N-1)(n-1)$ or 21 degrees of freedom. From a table of the chi-square distribution, Table 1.3,

$$\chi_{0.975}^2(21) = 35.48$$

$$\chi_{0.025}^2(21) = 10.28$$

Thus the 95% confidence limits on σ_0^2, in grams² UO₂, are

$$\frac{(21)(93.30)}{35.48} < \sigma_0^2 < \frac{(21)(93.30)}{10.18}$$

$$55.22 < \sigma_0^2 < 190.59$$

The corresponding 95% confidence limits on σ_0, in grams UO_2, are

$$7.43 < \sigma_0 < 13.81$$

Finally, test the hypothesis that the standard deviation is 5 grams. This might correspond to the design precision standard deviation for this type of scale. Thus σ_H^2 in Eq. 2.3.8 is 25 grams^2. For a significance level $p_1 = 0.01$, say,

$$\chi_{0.99}^2(21) = 38.96$$

the hypothesis H_0: $\sigma_0^2 = 25$ grams^2 is rejected if

$$M_0 > \frac{(25)(38.96)}{21}$$

or if

$$M_0 > 46.38$$

Since $M_0 = 93.30$, the hypothesis is rejected; the performance of the scales is not consistent with their design specification.   ∎

REFERENCES

1. Bennett, C. A. and Franklin, N. L. *Statistical Analysis in Chemistry and the Chemical Industry*. New York: John Wiley and Sons; 1954; p. 658.
2. Brownlee, K. A. *Statistical Theory and Methodology in Science and Engineering*. New York: John Wiley and Sons; 1960; Chapter 13.

Chapter 3

$N = 2$ MEASUREMENT METHODS

3.1 INTRODUCTION

In this chapter, the case of $N = 2$ measurement methods with $\sigma_1^2 \neq \sigma_2^2$ is studied with reference to the model given as Eq. 1.8.1. Further, $\beta_1 = \beta_2$ while $\alpha_1 \neq \alpha_2$ (constant bias model). Unlike Chapter 2, for $\sigma_1^2 \neq \sigma_2^2$ we now distinguish between random and fixed models. Such a distinction was not needed in Chapter 2, where only the differences between observations were studied, and the origin of the μ_k (true item) values did not play a role in making inferences about σ_δ^2 or σ_α^2.

Different estimators of σ_1^2 and of σ_2^2 are considered. In Section 3.2, the Grubbs' estimators are discussed, and in Section 3.3 these are compared with the maximum likelihood estimators (MLE's). Problems of statistical inference are considered in Section 3.4. In practice, the Grubbs' and MLE procedures may produce a negative estimate of σ_1^2 or σ_2^2; how to deal with this problem is the topic of Sections 3.5 and 3.6, Section 3.5 dealing with constrained maximum likelihood estimators (CMLE's) and Section 3.6 with constrained expected likelihood estimators (CELE's). The estimation of relative biases is discussed in Section 3.7.

3.2 GRUBBS' ESTIMATORS

Grubbs' estimators (1) are based on the moment method of estimation, in which sample moments are equated to population moments with the system of equations solved in some sense for the unknown parameters. For $N = 2$, the number of parameters is equal to the number of moment equations, and these are easily solved simultaneously to yield the estimates.

There are three unknown parameters: σ_1^2, σ_2^2, and σ_μ^2. For the random model, σ_μ^2 is defined by regarding μ_k as a normally distributed random variable with mean μ and variance σ_μ^2. For the fixed model, σ_μ^2 is defined by

$$\sigma_\mu^2 = \frac{\sum\limits_{k=1}^{n} (\mu_k - \mu)^2}{n - 1} \tag{3.2.1}$$

where μ is the mean of the n μ_k values.

With the Grubbs' estimation procedure, the estimates of σ_1^2, σ_2^2, and σ_μ^2 are the same for both the random and fixed models; the interpretation of σ_μ^2 is, of course, model dependent.

In finding Grubbs' estimates, the quantities s_1^2, s_2^2, and s_{12} are calculated, where s_i^2 is defined by Eq. 1.6.2 and s_{12} by Eq. 2.2.12.

For either the random or fixed model,

$$E\left(s_i^2\right) = \sigma_i^2 + \sigma_\mu^2; \qquad i = 1, 2$$

$$E\left(s_{12}\right) = \sigma_\mu^2 \tag{3.2.2}$$

It follows that the Grubbs' estimators are

$$\tilde{\sigma}_i^2 = s_i^2 - s_{12}; \qquad i = 1, 2 \tag{3.2.3}$$

$$\tilde{\sigma}_\mu^2 = s_{12} \tag{3.2.4}$$

The problems of finding interval estimates and of testing hypotheses about σ_1^2 and σ_2^2 are deferred until after the discussion on MLE's.

3.3 MAXIMUM LIKELIHOOD ESTIMATORS (MLE'S)

In his original article, Grubbs (1) states that, for the random model, his estimators are maximum likelihood. In a related article, Thompson (2) states that Grubbs' estimators have a maximum likelihood property. It is shown in the next section that Grubbs' estimators are identical to MLE's within the factor $(n - 1)/n$. In other words, $\tilde{\sigma}_i^2$ given by Eq. 3.2.3, when multiplied by $(n - 1)/n$, is the MLE. The same is true for $\tilde{\sigma}_\mu^2$. For the fixed model, this correspondence does not hold, as shown in Section 3.3.2.

3.3.1 Random Model

For the random model, the derivation of the MLE's is given in Appendix 3.A. It is shown that the following three equations in three unknowns result from applying MLE procedures:

$$\sigma_1^2 = \frac{(n-1)\left[\left(\sigma_\mu^2 + \sigma_2^2\right)^2 s_1^2 - 2\sigma_\mu^2\left(\sigma_\mu^2 + \sigma_2^2\right)s_{12} + \sigma_\mu^4 s_2^2\right]}{n\left(\sigma_\mu^2 + \sigma_2^2\right)^2} - \frac{\sigma_\mu^2 \sigma_2^2}{\sigma_\mu^2 + \sigma_2^2}$$

(3.3.1)

The equation for σ_2^2 is identical except that σ_1^2 and σ_2^2 are interchanged, as are s_1^2 and s_2^2.

$$\sigma_\mu^2 = \frac{(n-1)\left(\sigma_1^4 s_2^2 + 2\sigma_1^2\sigma_2^2 s_{12} + \sigma_2^4 s_1^2\right)}{n\left(\sigma_1^2 + \sigma_2^2\right)^2} - \frac{\sigma_1^2 \sigma_2^2}{\sigma_1^2 + \sigma_2^2}$$

(3.3.2)

The three equations appear formidable to solve. However, it is shown in Appendix 3.B that if the following equalities hold, then the three equations are satisfied:

$$\sigma_1^2 = \frac{(n-1)\left(s_1^2 - s_{12}\right)}{n}$$

$$\sigma_2^2 = \frac{(n-1)\left(s_2^2 - s_{12}\right)}{n}$$

(3.3.3)

$$\sigma_\mu^2 = \frac{(n-1)s_{12}}{n}$$

Since these values of the three parameters satisfy Eqs. 3.3.1 and 3.3.2, they are the MLE's, and are denoted by $\hat{\sigma}_1^2$, $\hat{\sigma}_2^2$, and $\hat{\sigma}_\mu^2$, respectively. Comparing the MLE's with the Grubbs' estimators in Eqs. 3.2.3 and 3.2.4 shows that they are identical within the $(n-1)/n$ factor.

Note that the Grubbs' estimates are unbiased, whereas the MLE's would have to be multiplied by $n/(n-1)$ in order to remove the bias. This is analogous to the simple problem of estimating a population variance σ^2 for a normally distributed random variable by s^2, the sum of the squared deviations from the sample mean divided by $(n-1)$, n being the sample size. The MLE of σ^2 is $(n-1)/n$ times the unbiased estimate s^2.

3.3.2 Fixed Model

For the fixed model, the derivation of the MLE's is given in Appendix 3.C. The logarithm of the likelihood function is

$$L = C - 0.5\left[n \ln \sigma_1^2 \sigma_2^2 + \frac{(n-1)\left(s_1^2 - 2s_{12} + s_2^2\right)}{\sigma_1^2 + \sigma_2^2}\right]$$

(3.3.4)

where C is a constant.

Note that L, being symmetric in σ_1^2 and σ_2^2, is a useless function. For example, the value of the likelihood with σ_1^2 and σ_2^2 equal to the Grubbs' estimates is exactly the same as that value when the two estimates are interchanged, a nonsensical result. Thus for the fixed model, one may obtain unbiased estimates of the parameters from Grubbs' Eqs. 3.2.3 and 3.2.4, but the estimates are unrelated to maximum likelihood.

Before the problems of statistical inference are addressed, some numerical examples are introduced to illustrate the results thus far.

EXAMPLE 3.A

An experiment was conducted in which the densities of 43 cylindrical nuclear reactor fuel pellets of sintered uranium were measured by 6 measurement methods. The experiment was described by Jaech (3), but the detailed data needed for this example were not given in the cited reference. The original 258 data points were not recoverable, but the data have been reconstructed to some extent in this example and in others to follow.

Table 3.1   Pellet Densities—90%

Method 1	Method 6	Method 1	Method 6
4.30	4.75	3.97	4.29
4.51	4.79	4.40	4.62
4.42	4.72	4.24	4.97
4.90	4.75	4.51	4.69
4.60	5.05	4.49	4.67
4.29	4.93	4.42	4.46
4.30	4.66	4.49	4.75
4.69	4.75	4.14	4.58
4.02	4.40	4.38	4.63
4.61	5.18	4.45	4.81
4.35	4.39	4.39	4.54
4.46	4.75	4.56	4.46
4.08	4.33	4.38	4.56
3.95	3.94	4.44	4.66
4.21	4.62	4.21	4.40
4.62	4.70	4.51	4.77
4.81	5.22	4.23	4.15
4.47	4.38	4.33	4.92
4.38	4.20	4.27	4.41
4.43	4.51	4.67	4.73
4.74	4.77		
4.35	4.54		
4.11	4.26		

Method 1 in this experiment was a geometric method for measuring density, based on weighing the pellet and finding its volume by measuring the pellet diameter and length. Method 6 was an immersion method, based on noting the change in weight when the pellet was weighed in both air and a liquid. Table 3.1 lists the data, with density expressed as percentage theoretical density (%TD) minus 90 for convenience.

From Eqs. 1.6.2 and 2.2.12, with the subscript 2 replaced by 6 to denote Method 6,

$$s_1^2 = 0.044954 \qquad s_6^2 = 0.069424 \qquad s_{16} = 0.034938$$

From Eqs. 3.2.3 and 3.2.4, the Grubbs' estimates of the parameters are

$$\tilde{\sigma}_1^2 = 0.010016 \qquad \tilde{\sigma}_6^2 = 0.034486 \qquad \tilde{\sigma}_\mu^2 = 0.034938$$

From Eq. 3.3.3, the MLE's are

$$\hat{\sigma}_1^2 = 0.009783 \qquad \hat{\sigma}_6^2 = 0.033684 \qquad \hat{\sigma}_\mu^2 = 0.034125$$

This example is continued following the presentation on statistical inference.   ∎

EXAMPLE 3.B

In a paper dealing with verification of irradiated fuel assemblies (4), Dowdy, Nicholson, and Caldwell list readings by three observers for the image "extinction" method of measuring Cerenkov glow intensity. Since the method requires the subjective judgment of the observer, it is subject to considerable error of measurement. The data for Observers 2 and 3 are reproduced as Table 3.2. From Eqs. 1.6.2 and 2.2.12 with subscripts 1 and 2 replaced by 2 and 3, respectively,

$$s_2^2 = 773.09 \times 10^{-10} \qquad s_3^2 = 555.11 \times 10^{-10} \qquad s_{23} = 113.52 \times 10^{-10}$$

From Eqs. 3.2.3 and 3.2.4, the Grubbs' estimates of the parameters are

$$\tilde{\sigma}_2^2 = 659.57 \times 10^{-10} \qquad \tilde{\sigma}_3^2 = 441.59 \times 10^{-10} \qquad \tilde{\sigma}_\mu^2 = 113.52 \times 10^{-10}$$

and from Eq. 3.3.3, the MLE's are

$$\hat{\sigma}_2^2 = 641.25 \times 10^{-10} \qquad \hat{\sigma}_3^2 = 429.32 \times 10^{-10} \qquad \hat{\sigma}_\mu^2 = 110.37 \times 10^{-10} \quad ∎$$

Table 3.2   Results (Burnup/Aperture Size) of the Image
             "Extinction" Method Used By Two
             Different Observers

Observer 2	Observer 3
1.64×10^{-3}	1.19×10^{-3}
1.56	1.20
1.26	1.20
1.18	1.18
1.31	1.31
1.47	1.68
1.15	1.03
1.44	1.44
1.44	1.67
1.72	1.72
1.72	1.49
1.51	1.51
1.91	1.53
2.07	1.55
0.95	1.04
1.63	1.93
2.08	1.16
1.90	1.16
1.89	1.49
1.89	1.49
1.44	1.98
1.70	2.02
1.50	1.50
1.45	1.60
1.49	1.50
1.31	1.31
1.27	1.46
1.27	1.45
1.55	1.55
1.25	1.61
1.21	1.56
1.34	1.54
1.75	1.48
1.76	1.49
1.91	1.36
1.75	1.38

3.4 INFERENCE

Statistical inference problems include both the construction of interval estimates and the testing of hypotheses.

3.4.1 Interval Estimation

The results given in this section all assume that the model is random.

Confidence Interval on σ_i^2. In his article, Grubbs (1) gives an expression for
the variance of $\tilde{\sigma}_i^2$ given by Eq. 3.2.3:

$$\operatorname{var} \tilde{\sigma}_i^2 = \frac{2\sigma_i^4}{n-1} + \frac{\sigma_\mu^2 \sigma_1^2 + \sigma_\mu^2 \sigma_2^2 + \sigma_1^2 \sigma_2^2}{n-1} \tag{3.4.1}$$

An evaluation of $\operatorname{var} \tilde{\sigma}_i^2$ requires one to know the values of the parameters,
which are, of course, unknown quantities. They may be replaced by their
estimates to estimate $\operatorname{var} \hat{\sigma}_i^2$.

Although Eq. 3.4.1 gives the sampling variance of $\tilde{\sigma}_i^2$, it does not indicate its
distribution. Under certain conditions, the precise nature of which has not
been investigated, one might validly assume that $\tilde{\sigma}_i^2$ is distributed as chi-square
with degrees of freedom, df, calculated by equating the variance of a chi-
squared distributed sampling variance to that of $\tilde{\sigma}_i^2$ and solving for df.
Specifically, equate

$$\frac{2\sigma_i^4}{df_i} = \operatorname{var} \tilde{\sigma}_i^2$$

and solve for df_i giving

$$df_i = \frac{2\tilde{\sigma}_i^4}{\operatorname{var} \tilde{\sigma}_i^2} \tag{3.4.2}$$

upon replacing σ_i^4 by its estimate. Once df_i is found, approximate $100(p_2 -
p_1)\%$ confidence limits on σ_i^2 may be found by solving for σ_i^2 the two-sided
inequality:

$$\chi_{p_1}^2(df_i) < \frac{(df_i)\tilde{\sigma}_i^2}{\sigma_i^2} < \chi_{p_2}^2(df_i) \tag{3.4.3}$$

where $\chi_{p_i}^2(df_i)$ is defined by Eq. 2.2.18 with ν degrees of freedom replaced by
df_i.

EXAMPLE 3.C

Example 3.A is continued by finding the approximate 95% confidence intervals
on σ_1^2 and σ_6^2. From Eq. 3.4.1, with the parameter values replaced by their

estimates,

$$\text{var}\,\tilde{\sigma}_1^2 = \left[2(0.010016)^2 + (0.034938)(0.010016) + (0.034938)(0.034886)\right.$$

$$\left. + (0.010016)(0.034886)\right]/42 = 0.5045 \times 10^{-4}$$

$$\text{var}\,\tilde{\sigma}_6^2 = 1.0363 \times 10^{-4}$$

The degrees of freedom are calculated from Eq. 3.4.2. For $i = 1$,

$$df_1 = \frac{2(0.010016)^2}{0.5045 \times 10^{-4}} = 3.98$$

For $i = 6$, $df_6 = 23.49$.

In applying Eq. 3.4.3 with $p_1 = 0.025$ and $p_2 = 0.975$, and interpolating in the chi-square from Table 1.2,

$$\chi^2_{0.025}(3.98) = 0.484 \qquad \chi^2_{0.975}(3.98) = 11.14$$

$$\chi^2_{0.025}(23.49) = 12.04 \qquad \chi^2_{0.975}(23.49) = 38.70$$

From Eq. 3.4.3 for $i = 1$,

$$0.484 < \frac{(3.98)(0.010016)}{\sigma_1^2} < 11.14$$

from which

$$0.003578 < \sigma_1^2 < 0.082363$$

is the approximate 95% confidence interval on σ_1^2. For $i = 6$,

$$12.04 < \frac{(23.49)(0.034886)}{\sigma_6^2} < 38.70$$

$$0.021175 < \sigma_6^2 < 0.068062 \qquad\qquad \blacksquare$$

EXAMPLE 3.D

Example 3.B is continued by finding the approximate 90% confidence intervals on σ_2^2 and σ_3^2. From Eq. 3.4.1 with the parameter values replaced by their estimates,

$$\text{var}\,\tilde{\sigma}_2^2 = \left[2(659.57)^2 + (113.52)(659.57) + (113.52)(441.59)\right.$$

$$\left. + (659.57)(441.59)\right] \times 10^{-20}/35$$

$$= 36{,}752 \times 10^{-20}$$

$$\text{var}\,\tilde{\sigma}_3^2 = 23{,}036 \times 10^{-20}$$

The degrees of freedom are calculated from Eq. 3.4.2. For $i = 2$,

$$df_2 = \frac{2(659.57)^2}{36,752} = 23.67$$

$$df_3 = \frac{2(441.59)^2}{23,036} = 16.93$$

In applying Eq. 3.4.3 with $p_1 = 0.050$ and $p_2 = 0.950$, and interpolating in the chi-square, Table 1.2,

$$\chi^2_{0.050}(23.67) = 13.60 \qquad \chi^2_{0.950}(23.67) = 36.01$$

$$\chi^2_{0.050}(16.93) = 8.62 \qquad \chi^2_{0.950}(16.93) = 27.50$$

From Eq. 3.4.3

$$13.60 < \frac{(23.67)(659.57) \times 10^{-10}}{\sigma_2^2} < 36.01$$

from which

$$433.55 \times 10^{-10} < \sigma_2^2 < 1147.94 \times 10^{-10}$$

is the approximate 90% confidence interval on σ_2^2. Also,

$$8.62 < \frac{(16.93)(441.59) \times 10^{-10}}{\sigma_3^2} < 27.50$$

$$271.86 \times 10^{-10} < \sigma_3^2 < 867.30 \times 10^{-10} \qquad ■$$

Simultaneous Confidence Region on σ_1^2, σ_2^2, and σ_μ^2. The method above for constructing a confidence interval for each parameter is an approximate one, and the approximation is probably quite poor for small n or large σ_μ^2 relative to the smaller of σ_1^2 and σ_2^2. Whatever the size of n and σ_μ^2, an exact method of constructing a simultaneous confidence region for the three parameters σ_1^2, σ_2^2, and σ_μ^2 is given by Thompson (2).

In our terminology, the probability is at least $(p_2 - p_1)$ that the following three relations hold simultaneously:

$$|\sigma_\mu^2 - (n-1)s_{12}K| \le M(n-1)s_1 s_2 \tag{3.4.4}$$

$$|\sigma_i^2 - (n-1)(s_i^2 - s_{12})K| \le M(n-1)s_i(s_1^2 + s_2^2 - 2s_{12})^{0.5}; \quad i = 1, 2 \tag{3.4.5}$$

Table 3.3   Constants Used to Construct Simultaneous Confidence Regions[a]

	$p_2 - p_1 = 0.99$		$p_2 - p_1 = 0.95$	
$n-1$	K	M	K	M
3	99.78	99.72	19.79	19.71
4	12.38	12.33	4.146	4.077
5	3.980	3.931	1.726	1.665
6	1.903	1.858	0.9636	0.9083
7	1.120	1.078	0.6290	0.5786
8	0.7459	0.7076	0.4516	0.4052
9	0.5389	0.5031	0.3453	0.3022
10	0.4120	0.3782	0.2761	0.2357
11	0.3282	0.2963	0.2280	0.1901
12	0.2698	0.2395	0.1932	0.1573
13	0.2272	0.1983	0.1668	0.1328
14	0.1951	0.1675	0.1464	0.1140
15	0.1702	0.1438	0.1301	0.09925
16	0.1505	0.1251	0.1169	0.08738
17	0.1344	0.1100	0.1060	0.07767
18	0.1213	0.09772	0.09682	0.06962
19	0.1103	0.08752	0.08904	0.06287
20	0.1009	0.07896	0.08237	0.05713
22	0.08610	0.06546	0.07152	0.04795
24	0.07484	0.05538	0.06311	0.04098
26	0.06605	0.04763	0.05641	0.03554
28	0.05901	0.04152	0.05096	0.03121
30	0.05328	0.03660	0.04644	0.02768
35	0.04272	0.02778	0.03796	0.02127
40	0.03556	0.02200	0.03205	0.01700
45	0.03040	0.01797	0.02771	0.01398
50	0.02652	0.01503	0.02440	0.01176
60	0.02109	0.01110	0.01967	0.00875
70	0.01748	0.00862	0.01646	0.00684
80	0.01492	0.00694	0.01415	0.00553
90	0.01300	0.00575	0.01241	0.00460
100	0.01152	0.00486	0.01104	0.00390

[a] Reprinted with permission of the American Statistical Association.

The parameters K and M were tabulated by Thompson for $(p_2 - p_1)$ equal to 0.99 and 0.95 and are included here as Table 3.3.

EXAMPLE 3.E

Consider Example 3.A and find the 95% simultaneous confidence region on σ_1^2, σ_6^2, and σ_μ^2. Here

$$n = 43 \qquad s_{16} = 0.034938$$

$$s_1^2 = 0.044954 \qquad s_6^2 = 0.069424$$

From Table 3.3, for $(n - 1) = 42$,

$$K = 0.03031 \qquad M = 0.01579$$

From Eq. 3.4.4,

$$|\sigma_\mu^2 - 0.04448| \leq 0.03705$$

which gives

$$0.00743 \leq \sigma_\mu^2 \leq 0.08153$$

From Eq 3.4.5,

$$|\sigma_1^2 - 0.01275| \leq 0.02966$$

and

$$|\sigma_6^2 - 0.04390| \leq 0.03686$$

from which

$$0 \leq \sigma_1^2 \leq 0.04241$$

and

$$0.00704 \leq \sigma_6^2 \leq 0.08076 \qquad\qquad \blacksquare$$

EXAMPLE 3.F

Construct a 95% simultaneous confidence region on σ_2^2, σ_3^2, and σ_μ^2 for the data of Example 3.B. From Example 3.B,

$$n = 36 \qquad s_{23} = 113.52 \times 10^{-10}$$

$$s_2^2 = 773.09 \times 10^{-10} \qquad s_3^2 = 555.11 \times 10^{-10}$$

From Table 3.3 for $(n - 1) = 35$,

$$K = 0.03796 \qquad M = 0.02127$$

From Eqs. 3.4.4 and 3.4.5,

$$|\sigma_\mu^2 - 150.8227 \times 10^{-10}| \leq 487.6858 \times 10^{-10}$$

$$|\sigma_2^2 - 876.3047 \times 10^{-10}| \leq 686.8719 \times 10^{-10}$$

$$|\sigma_3^2 - 586.6965 \times 10^{-10}| \leq 582.0366 \times 10^{-10}$$

from which the 95% confidence region is

$$0 \leq \sigma_\mu^2 \leq 638.51 \times 10^{-10}$$

$$189.43 \times 10^{-10} \leq \sigma_2^2 \leq 1563.18 \times 10^{-10}$$

$$4.66 \times 10^{-10} \leq \sigma_3^2 \leq 1168.73 \times 10^{-10} \qquad \blacksquare$$

3.4.2 Hypothesis Testing

There are a number of hypotheses about σ_1^2 and σ_2^2 that might be of interest. If the two measurement methods are basically the same analytical technique, but are performed perhaps by two different analysts or in two different laboratories, then the null hypothesis H_{01}: $\sigma_1^2 = \sigma_2^2$ would likely be of interest. An exact test of this hypothesis is discussed in this section. Hypothesized values may be specified for either or both of these parameters, and a number of different hypotheses may be of interest. The test of the null hypothesis H_{02}: $(\sigma_1^2 + \sigma_2^2) = \sigma_0^2$, with σ_0^2 specified, is also considered here. This is also an exact test. Finally, a number of large-sample tests are presented that may be used to test any number of different hypotheses about σ_1^2 or σ_2^2. Specifically, these are the test of

$$H_{03}: \quad \sigma_1^2 = \sigma_{10}^2, \qquad \text{with } \sigma_{10}^2 \text{ specified}$$

the test of

$$H_{04}: \quad \sigma_1^2 = \sigma_{10}^2 \qquad \text{and}$$

$$\sigma_2^2 = \sigma_{20}^2, \qquad \text{with } \sigma_{10}^2 \text{ and } \sigma_{20}^2 \text{ both specified}$$

and the test of

$$H_{05}: \quad \sigma_1^2 = k\sigma_2^2$$

For $k = 1$, H_{05} is the same as H_{01}.

Test of H_{01}: $\sigma_1^2 = \sigma_2^2$. The test of H_{01}: $\sigma_1^2 = \sigma_2^2$ is due to Maloney and Rastogi (5). Two new random variables, U and V, are introduced where

$$U = X_1 + X_2$$
$$V = X_1 - X_2$$
(3.4.6)

Then, two columns of numbers for the random variables U and V are formed by summing and differencing the pairs of data values, respectively. The sample variances of these sums (u_k) and differences (v_k) are then found using Eq. 1.6.2, and are denoted by s_u^2 and s_v^2, respectively. Next, the sample covariance between the u_k's and v_k's is calculated using Eq. 2.2.12 and denoted by s_{uv}. It is seen that

$$E(s_{uv}) = \sigma_1^2 - \sigma_2^2$$
(3.4.7)

The sample correlation coefficient between U and V is denoted by r_{uv} and calculated by

$$r_{uv} = \frac{s_{uv}}{s_u s_v}$$
(3.4.8)

Observe from Eq. 3.4.7 that under the null hypothesis H_{01}: $\sigma_1^2 = \sigma_2^2$, $E(s_{uv})$ is zero, and hence, a test of the hypothesis that the correlation coefficient between U and V is zero is equivalent to a test of H_{01}. As is mentioned in several texts (for example, Reference 6), an exact test on the sample correlation coefficient is due to R. A. Fisher, who pointed out that, under the null hypothesis that the correlation coefficient is zero,

$$t = \frac{r_{uv}\sqrt{n - 2}}{\sqrt{1 - r_{uv}^2}}$$
(3.4.9)

is distributed as Student's t-distribution with ($n - 2$) degrees of freedom.

In applying this test, one need not actually form the columns of sums and differences; in other words, u_k and v_k need not be calculated. Having already found s_1^2, s_2^2, and s_{12}, it follows from Eq. 3.4.6 that

$$s_u^2 = s_1^2 + s_2^2 + 2s_{12}$$

$$s_v^2 = s_1^2 + s_2^2 - 2s_{12}$$
(3.4.10)

$$s_{uv} = s_1^2 - s_2^2$$

Thus the test of H_{01}: $\sigma_1^2 = \sigma_2^2$ may be summarized as follows: Calculate s_u^2, s_v^2, and s_{uv} from Eq. 3.4.10, r_{uv} from Eq. 3.4.8, and t from Eq. 3.4.9. For a two-sided test, the alternative hypothesis being $\sigma_1^2 \neq \sigma_2^2$, and at significance level α, H_{01} is rejected if $|t|$ exceeds $t_p(\nu)$ defined by

$$\int_{-\infty}^{t_p(\nu)} t(\nu) \, dt = p \qquad\qquad (3.4.11)$$

that is, the area under the t-distribution curve with ν degrees of freedom is p. For this specific test, ν is $(n - 2)$ and p is $(1 - \alpha/2)$.

EXAMPLE 3.G

Example 3.A is continued. Recall that the MLE's of σ_1^2 and σ_6^2 were, respectively, 0.009783 and 0.033684, and that $n = 43$. Test H_{01}: $\sigma_1^2 = \sigma_6^2$ with $\alpha = 0.05$. From Example 3.A,

$$s_1^2 = 0.044954 \qquad s_6^2 = 0.069424 \qquad s_{16} = 0.034938$$

Then from Eqs. 3.4.10, 3.4.8, and 3.4.9,

$$s_u^2 = 0.184254$$

$$s_v^2 = 0.044502$$

$$s_{uv} = -0.024470$$

$$r_{uv} = -0.270$$

$$t = -1.796$$

From Table 1.3 with $\nu = 41$ degrees of freedom, and with $p = (1 - 0.05/2) = 0.975$, $t_p(\nu)$ is 2.020. Since $|t|$ does not exceed 2.020, the hypothesis of equal precisions is not rejected.   ∎

EXAMPLE 3.H

For the data of Example 3.B, test H_{01}: $\sigma_2^2 = \sigma_3^2$ with $\alpha = 0.05$. From Example 3.B,

$$s_2^2 = 773.09 \times 10^{-10} \qquad s_3^2 = 555.11 \times 10^{-10} \qquad s_{23} = 113.52 \times 10^{-10}$$

From Eqs. 3.4.10, 3.4.8, and 3.4.9,

$$s_u^2 = 1555.24 \times 10^{-10}$$

$$s_v^2 = 1101.16 \times 10^{-10}$$

$$s_{uv} = 217.98 \times 10^{-10}$$

$$r_{uv} = 0.167$$

$$t = 0.988$$

Since 0.988 is less than the critical value of 2.034 read from Table 1.3, H_{01} is not rejected. ∎

Test of H_{02}: $(\sigma_1^2 + \sigma_2^2) = \sigma_0^2$. For a given set of paired data, values may be hypothesized for σ_1^2 and σ_2^2. Although one's primary interest would be in testing H_{04}: $\sigma_1^2 = \sigma_{10}^2$ and $\sigma_2^2 = \sigma_{20}^2$, if the model is not random or if the sample size is not large enough to permit application of the large-sample test, discussed in what follows, then a test of H_{04} does not exist. However, an exact test of H_{02} for either a random or fixed model may be applied, which provides some insight into the truth of H_{04}. (Note that σ_{10}^2 could be overstated and σ_{20}^2 understated to the extent that H_{04} would be false but H_{02} true, so the two hypotheses are not equivalent.)

The test of H_{02} is identical to that given in Section 2.2.3, σ_0^2 replacing $2\sigma_H^2$. In Section 2.2.3, the test was set up as a one-sided test. The corresponding two-sided test with significance level α is to reject H_{02} if

$$s_d^2 < \frac{\sigma_0^2 \chi_{\alpha/2}^2 (n-1)}{n-1} \quad \text{or if} \quad s_d^2 > \frac{\sigma_0^2 \chi_{1-\alpha/2}^2 (n-1)}{n-1} \quad (3.4.12)$$

where the critical values are defined by Eq. 2.2.18.

EXAMPLE 3.I

Example 3.A is continued. Test the hypothesis H_{02}: $(\sigma_1^2 + \sigma_6^2) = 0.025$ for significance level $\alpha = 0.01$. Use a two-sided test of significance. For $n = 43$, the critical values are read from Table 1.2:

$$\chi_{0.005}^2 (42) = 22.13$$

$$\chi_{0.995}^2 (42) = 69.47$$

The hypothesis is rejected if s_d^2 is less than $(0.025)(22.13)/42$ or if it is greater than $(0.025)(69.47)/42$ or 0.0414. We note that s_d^2 is the same as s_v^2 defined earlier in this section and hence, from Example 3.G,

$$s_d^2 = 0.044502$$

Since $0.044502 > 0.0414$, the upper critical value, the hypothesis H_{02}: $(\sigma_1^2 + \sigma_6^2) = 0.025$, is rejected. ∎

Large-Sample Tests. The two tests discussed thus far have been exact tests, the test of H_{01} applying only to a random model and that of H_{02} to either a random or a fixed model. The large-sample tests considered in this section require that the model be random and that the sample size be "large," with the precise definition of large unspecified. It is not recommended that the large-sample tests discussed here and elsewhere in this book be applied for $n < 10$. For $10 \leq n \leq 20$, it is suggested that they be applied with caution, that is, by choosing a smaller value for α than would be chosen for $n > 20$.

The test statistic is the likelihood ratio. For the model under discussion, the logarithm of the likelihood, denoted by L, is given in Appendix 3.A as Eq. 3.A.10, with P defined in Eq. 3.A.5. To test any specified hypothesis about σ_1^2 and/or σ_2^2, the maximum value of L is found under that hypothesis and denoted by $L(\hat{\omega})$. Likewise, the maximum value of L is calculated in free-space, that is, upon inserting the MLE's of the parameters in Eq. 3.A.10. This value of L is denoted by $L(\hat{\Omega})$. Not counting α_1, α_2, and μ, which are estimated in both spaces Ω and ω, the number of parameters estimated in space Ω is three (σ_1^2, σ_2^2, and σ_μ^2). Letting the number of parameters estimated under the hypothesis, that is, in ω space, be r, then large-sample distribution theory tells us that

$$\lambda = -2[L(\hat{\omega}) - L(\hat{\Omega})] \qquad (3.4.13)$$

is approximately distributed as chi-square with $(3 - r)$ degrees of freedom. Note that in forming the difference $L(\hat{\omega}) - L(\hat{\Omega})$, the constant C drops out, so it need not be calculated.

In Appendix 3.D, it is shown that

$$L(\hat{\Omega}) = C - 0.5n \ln P - n \qquad (3.4.14)$$

where P is defined by Eq. 3.A.5 with σ_1^2, σ_2^2, and σ_μ^2 replaced by their MLE's, given by Eq. 3.3.3. The calculation of $L(\hat{\omega})$ depends on the stated hypothesis. In the following example, $L(\hat{\omega})$ is calculated for a number of hypotheses of potential interest.

EXAMPLE 3.J

This example will illustrate a test of the hypothesis H_{03}: $\sigma_1^2 = \sigma_{10}^2$. In Example 3.A as continued for $\alpha = 0.05$, test H_{03}: $\sigma_1^2 = 0.0225$ against the alternative hypothesis $\sigma_1^2 \neq 0.0225$. The values of σ_6^2 and σ_μ^2 are left unspecified in H_{03}.

First, $L(\hat{\Omega})$ is calculated from Eq. 3.4.14. From the cited example,

$$s_1^2 = 0.044954 \qquad s_6^2 = 0.069424 \qquad s_{16} = 0.034938$$

so that, from Eq. 3.D.1,

$$P = \frac{(42)^2 \left[(0.044954)(0.069424) - (0.034938)^2 \right]}{(43)^2}$$

$$= 0.001813$$

Then

$$L(\hat{\Omega}) = C - 21.5 \ln 0.001813 - 43$$

$$= C + 92.725$$

To find $L(\hat{\omega})$, L of Eq. 3.A.10 is maximized with respect to σ_6^2 and σ_μ^2 for $\sigma_1^2 = 0.0225$, the hypothesized value, and this maximum value of L is $L(\hat{\omega})$. The MLE's of σ_6^2 and σ_μ^2 corresponding to $\sigma_1^2 = 0.0225$ are the solutions to the following two equations, which are, respectively, Eqs. 3.3.1 and 3.3.2 with σ_1^2 and σ_2^2 interchanged and σ_2^2 replaced by σ_6^2, both with σ_1^2 set equal to 0.0225:

$$\sigma_6^2 = \frac{42 \left[(0.069424)\left(\sigma_\mu^2 + 0.0025\right)^2 - 0.069876\sigma_\mu^2\left(\sigma_\mu^2 + 0.0225\right) + 0.044954\sigma_\mu^4 \right]}{43\left(\sigma_\mu^2 + 0.0225\right)^2}$$

$$- \frac{0.0225\sigma_\mu^2}{\sigma_\mu^2 + 0.0225}$$

$$\sigma_\mu^2 = \frac{42 \left[(0.0225)^2(0.069424) + 2(0.0225)(0.034938)\sigma_6^2 + 0.0225\sigma_6^4 \right]}{43\left(0.0225 + \sigma_6^2\right)^2}$$

$$- \frac{0.0225\sigma_6^2}{0.0225 + \sigma_6^2}$$

These equations can be solved by trial and error. An initial value is assigned to σ_6^2, and σ_μ^2 is calculated from the second equation. This value is inserted in the first equation, and a new value of σ_6^2 is found. The iterative process is repeated until convergence occurs.

The MLE of σ_6^2 in freespace was 0.0337. This is used as the initial value in the iterative estimation procedure. The following sequences of values for σ_μ^2 and σ_6^2 are calculated:

Iteration	σ_6^2	σ_μ^2
1	0.033700	0.021664
2	0.033858	0.021592
3	0.033898	0.021574
4	0.033908	0.021570
5	0.033910	0.021569
6	0.033911	0.021568

Therefore, the estimates of σ_6^2 and σ_μ^2 in ω space corresponding to the stated hypothesis that $\sigma_1^2 = 0.0225$ are 0.033911 and 0.021568, respectively. These values are now inserted in Eq. 3.A.10 to calculate $L(\hat{\omega})$. First, in ω space from Eq. 3.A.5,

$$P = (0.021568)(0.0225) + (0.021568)(0.033911)$$

$$+ (0.0225)(0.033911) = 0.001980$$

Then,

$$L(\hat{\omega}) = C - 21.5\ln(0.001980)$$

$$-\frac{21[(0.055479)(0.044954) - (0.043136)(0.034938) + (0.044068)(0.069424)]}{0.001980}$$

$$= C + 90.915$$

Therefore, from Eq. 3.4.13,

$$\lambda = -2(90.915 - 92.725)$$

$$= 3.620$$

There is $(3 - 2)$ or 1 degree of freedom, since there were $r = 2$ parameters estimated in ω space. From Table 1.2, at $\alpha = 0.05$, the hypothesis will be rejected if $\lambda > 3.84$. Since $3.620 < 3.84$, H_{03}: $\sigma_1^2 = 0.0225$ is not rejected. ■

EXAMPLE 3.K

This example illustrates a test of the hypothesis H_{04}: $\sigma_1^2 = \sigma_{10}^2$ and $\sigma_6^2 = \sigma_{60}^2$. Continuing with Example 3.A, recall that in Example 3.I the hypothesis

$(\sigma_1^2 + \sigma_6^2) = 0.025$ was rejected for $\alpha = 0.01$. Let us now test the joint hypothesis $\sigma_1^2 = 0.010$ and $\sigma_2^2 = 0.015$, again for $\alpha = 0.01$. The value for $L(\hat{\Omega})$ is the same as in Example 3.J, namely, $L(\hat{\Omega}) = C + 92.725$.

To find $L(\hat{\omega})$, first find the MLE of σ_μ^2 corresponding to $\sigma_1^2 = 0.010$ and $\sigma_6^2 = 0.015$. This is found from Eq. 3.3.2 with 0.010 and 0.015 replacing σ_1^2 and σ_6^2, respectively, where σ_2^2 in Eq. 3.3.2 becomes σ_6^2 in this example.

$$\sigma_\mu^2 = \frac{42[(0.0001)(0.069424) + (0.069876)(0.00015) + (0.000225)(0.044954)]}{43(0.000625)}$$

$$-\frac{0.00015}{0.025}$$

$$= 0.037037$$

This, together with $\sigma_1^2 = 0.010$ and $\sigma_6^2 = 0.015$, is inserted into Eq. 3.A.10 to calculate $L(\hat{\omega})$. First, from Eq. 3.A.5,

$$P = (0.037037)(0.010) + (0.037037)(0.015) + (0.010)(0.015)$$

$$= 0.001076$$

Then,

$$L(\hat{\omega}) = C - 21.5 \ln (0.001076)$$

$$-\frac{21[(0.052037)(0.044954) - (0.069876)(0.037037) + (0.047037)(0.069424)]}{(0.001076)}$$

$$= C + 88.064$$

Therefore, from Eq. 3.4.13,

$$\lambda = -2(88.064 - 92.725) = 9.322$$

There are now $(3 - 1)$ or 2 degrees of freedom, since only one parameter was estimated in ω space. From Table 1.2, at $\alpha = 0.01$, the hypothesis is rejected if $\lambda > 9.21$. Since $9.322 > 9.21$, the hypothesis H_{04}: $\sigma_1^2 = 0.010$ and $\sigma_6^2 = 0.015$ is rejected at the $= 0.01$ level. In view of Example 3.I, we would expect that H_{04} would be rejected, since we have already rejected the hypothesis that $\sigma_1^2 + \sigma_6^2 = 0.025$. The hypothesis H_{04} is the more restrictive hypothesis. ∎

EXAMPLE 3.L

This example will illustrate a test of the hypothesis H_{05}: $\sigma_1^2 = k\sigma_6^2$. Continuing with Example 3.A, test the null hypothesis H_{05}: $\sigma_1^2 = 0.10\sigma_6^2$ against the alternative hypothesis that $k \neq 0.10$, for $\alpha = 0.05$.

The value for $L(\hat{\Omega})$ is the same as in Example 3.J:

$$L(\hat{\Omega}) = C + 92.725$$

To find $L(\hat{\omega})$, use is made of the MLE's of σ_6^2 and σ_μ^2 corresponding to $\sigma_1^2 = 0.10\sigma_6^2$. These MLE's are given by Jaech (7):

$$\sigma_2^2 = \frac{(n-1)(s_1^2 - 2s_{12} + s_2^2)}{n(k+1)} \qquad (3.4.15)$$

$$\sigma_\mu^2 = \frac{(n-1)[(1-k)(s_1^2 - ks_2^2) + 4ks_{12}]}{n(k+1)^2} \qquad (3.4.16)$$

For this example, replacing σ_2^2 by σ_6^2,

$$\hat{\sigma}_6^2 = \frac{42(0.044502)}{43(1.10)} = 0.039516$$

$$\hat{\sigma}_\mu^2 = \frac{42[(0.90)(0.038012) + 0.013975]}{43(1.10)^2} = 0.038897$$

From Eq. 3.A.5,

$$P = (0.038897)(0.003952) + (0.038897)(0.039516) + (0.003952)(0.039516)$$

$$= 0.001847$$

Then,

$$L(\hat{\omega}) = C - 21.5\ln(0.001847)$$

$$-\frac{21[(0.078413)(0.044954) - (0.069876)(0.038897) + (0.042849)(0.069424)]}{0.001847}$$

$$= C + 92.327$$

Therefore, from Eq. 3.4.13,

$$\lambda = -2(92.327 - 92.725) = 0.796$$

From Example 3.J, the critical value is 3.84. Therefore, the null hypothesis, $\sigma_1^2 = 0.10\sigma_6^2$, is not rejected.   ■

3.5   CONSTRAINED MAXIMUM LIKELIHOOD ESTIMATION (CMLE)

One of the σ_i^2 estimates given either by Grubbs in Eq. 3.2.3 or, for the random model, by the MLE of Eq. 3.3.3 may, in practice, be a negative quantity. This will occur with high probability if σ_μ^2 is large relative to σ_1^2 or σ_2^2. To provide nonnegative estimates of the parameters, constrained maximum likelihood estimators (CMLE's) are given in Section 3.5.1 for the random model in the event σ_i^2, for $i = 1$ or 2, is negative and in Section 3.5.2 in the event $\tilde{\sigma}_\mu^2$ of Eq. 3.2.4 is negative. The estimators discussed here are due to Thompson (8), who indicates that if $\hat{\sigma}_i^2$ is negative, then the MLE of σ_i^2 is zero.

3.5.1   Negative Estimate of Precision

Without loss of generality, assume that $\hat{\sigma}_1^2$ of Eq. 3.3.3 is negative. Note that both $\hat{\sigma}_1^2$ and $\hat{\sigma}_2^2$ cannot be negative. In Appendix 3.E, the derivation is given of the following MLE's of σ_2^2 and σ_μ^2 with σ_1^2 constrained to equal zero:

$$\hat{\sigma}_2^2 = \frac{(n-1)(s_1^2 - 2s_{12} + s_2^2)}{n} \tag{3.5.1}$$

$$\hat{\sigma}_\mu^2 = \frac{(n-1)s_1^2}{n} \tag{3.5.2}$$

These are called *constrained maximum likelihood estimates* (CMLE's).

3.5.2   Negative Estimate of Process Variance

In the event that $\hat{\sigma}_\mu^2$ of Eq. 3.3.3 is negative, the following MLE's of σ_1^2 and σ_2^2 are derived in Appendix 3.F:

$$\hat{\sigma}_i^2 = \frac{(n-1)s_i^2}{n}; \qquad i = 1, 2 \tag{3.5.3}$$

It is noted from Eqs. 3.5.1 to 3.5.3 that problems of statistical inference for CMLE's are straightforward. All estimates are directly related to sample variances that have chi-square density functions. Note in Eq. 3.5.1 that $(s_1^2 - 2s_{12} + s_2^2)$ is identical to s_d^2, as discussed in Section 2.2.3.

Table 3.4   Fuse Burning Times
(seconds)

Observer 1	Observer 2
10.10	10.07
9.98	9.90
9.89	9.85
9.79	9.71
9.67	9.65
9.89	9.83
9.82	9.75
9.59	9.56
9.76	9.68
9.93	9.89
9.62	9.61
10.24	10.23
9.84	9.83
9.62	9.58
9.60	9.60
9.74	9.73
10.32	10.32
9.86	9.86
9.65	9.64
9.50	9.49
9.56	9.56
9.54	9.53
9.89	9.89
9.53	9.52
9.52	9.52
9.44	9.43
9.67	9.67
9.77	9.76
9.86	9.84

EXAMPLE 3.M

Grubbs (1) gives an example in which fuse-burning times are reported by three observers. Restricting the analysis in this example to the data for the first two observers, the data are shown in Table 3.4. From Eqs. 1.6.2 and 2.2.12,

$$n = 29 \qquad s_1^2 = 0.046754 \qquad s_2^2 = 0.045112 \qquad s_{12} = 0.045582$$

From Eq. 3.3.3, the MLE's are

$$\hat{\sigma}_1^2 = 0.001132 \qquad \hat{\sigma}_2^2 = -0.000454 \qquad \hat{\sigma}_\mu^2 = 0.044010$$

Since $\hat{\sigma}_2^2$ is negative, the estimate of σ_2^2 is zero, while from Eqs. 3.5.1 and 3.5.2, the CMLE's of σ_1^2 and σ_μ^2 are

$$\hat{\sigma}_1^2 = 0.000678 \qquad \hat{\sigma}_\mu^2 = 0.045142 \qquad \blacksquare$$

EXAMPLE 3.N

Cylinders of UF_6 gas are shipped from a uranium enrichment facility to a reactor fuel fabrication facility. Samples of the gas are measured for percentage uranium by the shipper and by the receiver. For 26 paired sample measurements made over several consecutive shipments, the following are the summarized input statistics with all variances in percentage uranium squared:

$$n = 26 \quad s_1^2 = 0.635385 \times 10^{-4} \qquad s_2^2 = 0.481538 \times 10^{-4}$$

$$s_{12} = -0.049231 \times 10^{-4}$$

Since s_{12}, and hence $\hat{\sigma}_\mu^2$ of Eq. 3.3.3 are negative, the CMLE's of σ_1^2 and σ_2^2 are given by Eq. 3.5.3. The estimate of σ_μ^2 is zero.

$$\hat{\sigma}_1^2 = 0.610947 \times 10^{-4} \qquad \hat{\sigma}_2^2 = 0.463017 \times 10^{-4} \qquad \blacksquare$$

3.6   CONSTRAINED EXPECTED LIKELIHOOD ESTIMATION (CELE)

Although the CMLE's resolve the problem of negative estimates, the estimate of zero for one of the parameters remains a problem. Suppose that one would like to obtain positive estimates of both σ_1^2 and σ_2^2, as compared to simply nonnegative estimates. Such positive estimates may be derived by finding the expected likelihood estimates of σ_1^2 and σ_2^2, where the expectation is restricted to the space of nonnegative values for both parameters. Jaech (9) calls the resulting estimates *constrained expected likelihood estimates* (CELE's). Again, the random model applies.

The CELE's are related to the MLE's and CMLE's as follows. The likelihood function L of Eq. 3.A.10 provides the basis for all estimators. In finding the MLE's, L is maximized with respect to the parameters over all values of these parameters; in finding the CMLE's, L is maximized over the space of nonnegative values of the parameters. For the CELE's, attention is again restricted to the space of nonnegative values, but $\hat{\sigma}_i^2$ ($i = 1, 2$) is the *expected* value of σ_i^2 over this restricted space, and not the *maximum* value.

It is shown in Appendix 3.G that the CELE of σ_1^2, denoted by $\dot\sigma_1^2$, is

$$\dot\sigma_1^2 = \frac{SI_0}{I_1} \tag{3.6.1}$$

where

$$S = \frac{(n-1)\left(s_1^2 + s_2^2 - 2s_{12}\right)}{n} \tag{3.6.2}$$

$$I_0 = \int_0^1 vf(v)\,dv \tag{3.6.3}$$

$$I_1 = \int_0^1 f(v)\,dv \tag{3.6.4}$$

$$f(v) = \left[s_1^2(1-v)^2 + v^2 s_2^2 + 2v(1-v)s_{12}\right]^{-0.5n}$$

$$= \left[\frac{nSv^2}{n-1} + 2\left(s_{12} - s_1^2\right)v + s_1^2\right]^{-0.5n} \tag{3.6.5}$$

The integrals I_0 and I_1 are easily evaluated by numerical integration, as will be illustrated.

Once $\dot\sigma_1^2$ is computed, then the CELE of σ_2^2 is

$$\dot\sigma_2^2 = S - \dot\sigma_1^2 \tag{3.6.6}$$

and the CELE of σ_μ^2 is computed from Eq. 3.3.2 with σ_1^2 and σ_2^2 replaced by $\dot\sigma_1^2$ and $\dot\sigma_2^2$, respectively.

EXAMPLE 3.O

The CELE's are found for the data of Example 3.M. Recall that in Example 3.M, the MLE of σ_2^2 was negative. The relevant statistics were

$$n = 29 \qquad s_{12} = 0.045582$$

$$s_1^2 = 0.046754 \qquad s_2^2 = 0.045112$$

Applying Eqs. 3.6.2 and 3.6.5,

$$S = 0.000678 \left(\text{the CMLE of } \sigma_1^2 \text{ in the cited example}\right)$$

$$f(v) = \left(0.000702v^2 - 0.002344v + 0.046754\right)^{-14.5}$$

Table 3.5   CELE's for Example 3.0

Number of Intervals	Rule	I_0	I_1	$\dot{\sigma}_1^2 = I_0/I_1$
10	Trapezoidal	9.6969×10^{15}	2.6326×10^{19}	0.000368
10	Simpson's	9.6839×10^{15}	2.6330×10^{19}	0.000368
20	Trapezoidal	9.6871×10^{15}	2.6329×10^{19}	0.000368
20	Simpson's	9.6838×10^{15}	2.6330×10^{19}	0.000368

Then, I_0 and I_1 are evaluated by numerical integration, using the trapezoidal rule and Simpson's rule, with 10 and 20 intervals in each instance to show the comparison. The results are summarized in Table 3.5. Thus

$$\dot{\sigma}_1^2 = 0.000368$$

$$\dot{\sigma}_2^2 = 0.000678 - 0.000368 = 0.000310$$

From Eq. 3.3.2,

$$\dot{\sigma}_\mu^2 = 0.043945 \qquad \blacksquare$$

EXAMPLE 3.P

In the event s_{12} is smaller than both s_1^2 and s_2^2, the CELE's will tend to agree closely with the MLE's. This is illustrated for the Example 3.A data, for which

$$n = 43 \qquad\qquad s_{16} = 0.034938$$

$$s_1^2 = 0.044954 \qquad s_6^2 = 0.069424$$

The MLE's were

$$\hat{\sigma}_1^2 = 0.009783 \qquad \hat{\sigma}_6^2 = 0.033684 \qquad \hat{\sigma}_\mu^2 = 0.034125$$

Applying Eqs. 3.6.4 and 3.6.5:

$$S = 0.043467$$

$$f(v) = (0.044502v^2 - 0.020032v + 0.044954)^{-21.5}$$

For 10 intervals,

$$I_0 = 1.0616 \times 10^{27} \text{ by the trapezoidal rule}$$

$$= 1.0644 \times 10^{27} \text{ by Simpson's rule}$$

$$I_1 = 9.7999 \times 10^{28} \text{ by the trapezoidal rule}$$

$$= 9.8743 \times 10^{28} \text{ by Simpson's rule}$$

By both numerical integration methods, from Eq. 3.6.1,

$$\dot{\sigma}_1^2 = 0.0108$$

$$\dot{\sigma}_6^2 = 0.0327$$

From Eq. 3.3.2,

$$\dot{\sigma}_\mu^2 = 0.0336 \qquad\blacksquare$$

3.7   ESTIMATION OF RELATIVE BIASES

The development in Section 2.2.4 for the common precisions model is generally applicable here also. As in Section 2.2.4, one cannot estimate both α_1 and α_2, but only their difference, $(\alpha_1 - \alpha_2)$, which is estimated by $\bar{d}$. Specifically,

$$\widehat{\alpha_1 - \alpha_2} = \bar{d} = \bar{x}_1 - \bar{x}_2 \qquad (3.7.1)$$

For the case in which α_1 and α_2 are randomly selected from a population of α's with zero mean and variance σ_α^2, Eq. 2.2.24 is rewritten

$$E(\bar{d}^2) = 2\sigma_\alpha^2 + \frac{\sigma_1^2 + \sigma_2^2}{n} \qquad (3.7.2)$$

so that σ_α^2 is estimated by

$$\hat{\sigma}_\alpha^2 = 0.5\bar{d}^2 - \frac{0.5(\hat{\sigma}_1^2 + \hat{\sigma}_2^2)}{n} \qquad (3.7.3)$$

EXAMPLE 3.Q

In Example 3.A, in which reactor fuel pellet densities were measured by two methods, calculations show that

$$\bar{x}_1 = 4.397$$

$$\bar{x}_6 = 4.620$$

$$\bar{x}_1 - \bar{x}_6 = -0.223$$

Since the two measurement methods are quite different in this example, Eq. 3.7.1 would be more applicable than Eq. 3.7.3. Thus $(\alpha_1 - \alpha_2)$ is estimated to be -0.223% theoretical density.   ■

EXAMPLE 3.R

In Example 3.M, the average difference between the observers is

$$\bar{d} = \bar{x}_1 - \bar{x}_2 = 0.0238 \text{ sec}$$

In applying Eq. 3.7.3 to estimate σ_α^2, it does not matter whether one chooses to use the MLE's of σ_1^2 and σ_2^2, the CMLE's, or the CELE's; for any of these, the two estimates give

$$\hat{\sigma}_1^2 + \hat{\sigma}_2^2 = 0.000859$$

Therefore, from Eq. 3.7.3 with $n = 29$, σ_α^2 is estimated to be

$$\hat{\sigma}_\alpha^2 = 0.000283 - 0.000015$$

$$= 0.000268 \text{ sec}^2 \qquad\qquad ■$$

DERIVATION OF MAXIMUM LIKELIHOOD ESTIMATES FOR RANDOM MODEL

Consider the kth pair of observations (x_{1k}, x_{2k}). From Eq. 1.8.1, $i = 1$, with $\beta_i = 1$,

$$E(x_{ik}) = \alpha_i + \mu; \qquad i = 1, 2 \qquad (3.A.1)$$

$$E(x_{ik} - \alpha_i - \mu)^2 = \sigma_\mu^2 + \sigma_i^2; \qquad i = 1, 2 \qquad (3.A.2)$$

$$E(x_{1k} - \alpha_1 - \mu)(x_{2k} - \alpha_2 - \mu) = \sigma_\mu^2 \qquad (3.A.3)$$

Thus the variance-covariance matrix is

$$(\sigma_{ij}) = \begin{pmatrix} \sigma_\mu^2 + \sigma_1^2 & \sigma_\mu^2 \\ \sigma_\mu^2 & \sigma_\mu^2 + \sigma_2^2 \end{pmatrix} \qquad (3.A.4)$$

The determinant of (σ_{ij}), denoted by P, is

$$P = \sigma_\mu^2 \sigma_1 + \sigma_\mu^2 \sigma_2^2 + \sigma_1^2 \sigma_2^2 \qquad (3.A.5)$$

and the inverse matrix is

$$(\sigma^{ij}) = \frac{1}{P} \begin{pmatrix} \sigma_\mu^2 + \sigma_2^2 & -\sigma_\mu^2 \\ -\sigma_\mu^2 & \sigma_\mu^2 + \sigma_1^2 \end{pmatrix} \qquad (3.A.6)$$

The likelihood for the kth observation may now be written, its density function being the bivariate normal density:

$$f(x_{1k}, x_{2k}) = \frac{1}{2\pi\sqrt{P}} \exp\left\{ \frac{-0.5}{P} \left[(\sigma_\mu^2 + \sigma_2^2)(x_{1k} - \alpha_1 - \mu)^2 \right.\right.$$

$$- 2\sigma_\mu^2 (x_{1k} - \alpha_1 - \mu)(x_{2k} - \alpha_2 - \mu)$$

$$\left.\left. + (\sigma_\mu^2 + \sigma_1^2)(x_{2k} - \alpha_2 - \mu)^2 \right] \right\} \qquad (3.A.7)$$

For the n observations, the likelihood is found by raising the coefficient $(2\pi\sqrt{P})^{-1}$ to the nth power and summing the exponent for $k = 1$ to n. The natural logarithm of the likelihood function, denoted by L, is then

$$L = -n \ln 2\pi - 0.5n \ln P - 0.5P^{-1}$$

$$\times \left[\left(\sigma_\mu^2 + \sigma_2^2 \right) \sum_{k=1}^n (x_{1k} - \alpha_1 - \mu)^2 \right.$$

$$- 2\sigma_\mu^2 \sum_{k=1}^n (x_{1k} - \alpha_1 - \mu)(x_{2k} - \alpha_2 - \mu)$$

$$\left. + \left(\sigma_\mu^2 + \sigma_1^2 \right) \sum_{k=1}^n (x_{2k} - \alpha_2 - \mu)^2 \right] \qquad (3.A.8)$$

One cannot obtain separate estimates of the parameters α_1, α_2, and μ, but the sums $(\alpha_1 + \mu)$ and $(\alpha_2 + \mu)$ [and hence the difference $(\alpha_1 - \alpha_2)$] can be estimated. For $i = 1$ and 2, it follows easily upon differentiating L with respect to $(\alpha_i + \mu)$ that the MLE of $(\alpha_i + \mu)$ is

$$\widehat{\alpha_i + \mu} = \frac{\sum_{k=1}^n x_{ik}}{n} = \bar{x}_i; \qquad i = 1, 2 \qquad (3.A.9)$$

Upon replacing $(\alpha_i + \mu)$ in Eq. 3.A.8 by $\bar{x}_i$ for $i = 1$ and 2 and applying Eqs. 1.6.2 and 2.2.12, L may be rewritten

$$L = C - 0.5n \ln P - 0.5(n-1)P^{-1}\left[\left(\sigma_\mu^2 + \sigma_2^2 \right)s_1^2 - 2\sigma_\mu^2 s_{12} + \left(\sigma_\mu^2 + \sigma_1^2 \right)s_2^2 \right]$$

$$(3.A.10)$$

where $C = -n \ln 2\pi$, a constant.

The partial derivative of L with respect to σ_1^2 is found and equated to zero, keeping in mind that P is also a function of σ_1^2 and that

$$\frac{\partial P}{\partial \sigma_1^2} = \sigma_\mu^2 + \sigma_2^2 \qquad (3.A.11)$$

$$\frac{\partial L}{\partial \sigma_1^2} = \frac{-0.5n\left(\sigma_\mu^2 + \sigma_2^2 \right)}{P} - \frac{0.5(n-1)(\text{NUM})_1}{P^2} = 0 \qquad (3.A.12)$$

where

$$(\text{NUM})_1 = -\left[\left(\sigma_\mu^2 + \sigma_2^2\right)^2 s_1^2 - 2\sigma_\mu^2\left(\sigma_\mu^2 + \sigma_2^2\right)s_{12} + \sigma_\mu^4 s_2^2\right] \quad (3.A.13)$$

Upon multiplying both sides of Eq. 3.A.12 by $-2P^2$ and solving for σ_1^2, the solution follows easily:

$$\sigma_1^2 = \frac{-(n-1)(\text{NUM})_1}{n\left(\sigma_\mu^2 + \sigma_2^2\right)^2} - \frac{\sigma_\mu^2\sigma_2^2}{\sigma_\mu^2 + \sigma_2^2} \quad (3.A.14)$$

which is Eq. 3.3.1. The equation for σ_2^2 is easily found by symmetry. Finally, the partial derivative of L with respect to σ_μ^2 is equated to zero:

$$\frac{\partial L}{\partial\sigma_\mu^2} = \frac{-0.5n\left(\sigma_1^2 + \sigma_2^2\right)}{P} - \frac{0.5(n-1)(\text{NUM})_2}{P^2} = 0 \quad (3.A.15)$$

where

$$(\text{NUM})_2 = -\left(\sigma_1^4 s_2^2 + 2\sigma_1^2\sigma_2^2 s_{12} + \sigma_2^4 s_1^2\right) \quad (3.A.16)$$

The solution for σ_μ^2 is

$$\hat{\sigma}_\mu^2 = \frac{-(n-1)(\text{NUM})_2}{n\left(\sigma_1^2 + \sigma_2^2\right)^2} - \frac{\sigma_1^2\sigma_2^2}{\sigma_1^2 + \sigma_2^2} \quad (3.A.17)$$

which is Eq. 3.3.2.

PROOF THAT THE MAXIMUM LIKELIHOOD ESTIMATES ARE $(n - 1)/n$ TIMES THE GRUBBS' ESTIMATORS FOR RANDOM MODEL

In Eq. 3.3.1, replace σ_1^2, σ_2^2, and σ_μ^2 as indicated in Eq. 3.3.3 and demonstrate that the equality holds. The left-hand side (LHS) of Eq. 3.3.1 is

$$\text{LHS} = \frac{(n - 1)(s_1^2 - s_{12})}{n} \tag{3.B.1}$$

The right-hand side (RHS) is simplified by noting that

$$\sigma_\mu^2 + \sigma_2^2 = \frac{(n - 1)s_2^2}{n} \tag{3.B.2}$$

Then

$$\text{RHS} = \frac{n^2(n - 1)^3(s_1^2 s_2^4 - 2s_2^2 s_{12}^2 + s_2^2 s_{12}^2)}{n^3(n - 1)^2 s_2^4} - \frac{n(n - 1)^2 s_{12}(s_2^2 - s_{12})}{n^2(n - 1)s_2^2}$$

$$= \frac{n - 1}{n}\left(s_1^2 - \frac{s_{12}^2}{s_2^2} - s_{12} + \frac{s_{12}^2}{s_2^2} \right)$$

$$= \frac{(n - 1)(s_1^2 - s_{12})}{n} \tag{3.B.3}$$

which is identical to the LHS.

Similarly, if σ_1^2, σ_2^2, and σ_μ^2 are replaced in Eq. 3.3.2 by the quantities indicated in Eq. 3.3.3, the equality in Eq. 3.3.2 would also be shown to hold. Therefore, it has been proved that for this model the MLE's are $(n - 1)/n$ times the Grubbs' estimators.

DERIVATION OF MAXIMUM LIKELIHOOD ESTIMATES FOR FIXED MODEL

From Eq. 1.8.1 with $\beta_i = 1$,

$$E(x_{ik}) = \alpha_i + \mu_k; \qquad i = 1, 2 \quad (3.C.1)$$

$$E(x_{ik} - \alpha_i - \mu_k)^2 = \sigma_i^2; \qquad i = 1, 2 \quad (3.C.2)$$

$$E(x_{1k} - \alpha_1 - \mu_k)(x_{2k} - \alpha_2 - \mu_k) = 0 \qquad (3.C.3)$$

The variance-covariance matrix is

$$(\sigma_{ij}) = \begin{pmatrix} \sigma_1^2 & 0 \\ 0 & \sigma_2^2 \end{pmatrix} \qquad (3.C.4)$$

This leads to a simple expression for the log likelihood function for the n observations:

$$L = C - 0.5n \ln \sigma_1^2 \sigma_2^2 - 0.5 \left(\frac{\sum_{k=1}^{n} (x_{1k} - \alpha_1 - \mu_k)^2}{\sigma_1^2} + \frac{\sum_{k=1}^{n} (x_{2k} - \alpha_2 - \mu_k)^2}{\sigma_2^2} \right)$$

$$(3.C.5)$$

There are $(n + 4)$ unknown parameters; the n μ_k values, α_1, α_2, σ_1^2, and σ_2^2. The log likelihood L is differentiated with respect to each of these parameters, the partial derivatives are equated to zero, and the $(n + 4)$ equations are solved simultaneously. The solution for σ_i^2 is

$$\hat{\sigma}_i^2 = \frac{\sum\limits_{k=1}^{n} (x_{ik} - \alpha_i - \mu_k)^2}{n} \; ; \qquad i = 1, 2 \qquad (3.C.6)$$

Also,

$$\hat{\alpha}_i = \bar{x}_i = \frac{\sum\limits_{k=1}^{n} \mu_k}{n} \; ; \qquad i = 1, 2 \qquad (3.C.7)$$

and

$$\hat{\mu}_k = \frac{\sigma_2^2(x_{1k} - \alpha_1) + \sigma_1^2(x_{2k} - \alpha_2)}{\sigma_1^2 + \sigma_2^2} \; ; \qquad k = 1, 2, \ldots, n \qquad (3.C.8)$$

In solving these three sets of equations simultaneously, upon replacing α_i and μ_k by $\hat{\alpha}_i$ and $\hat{\mu}_k$, the expression $\sum_{k=1}^{n}(x_{ik} - \alpha_i - \mu_k)^2$ becomes

$$\frac{(n-1)\sigma_i^4(s_1^2 - 2s_{12} + s_2^2)}{(\sigma_1^2 + \sigma_2^2)^2} \; ; \qquad i = 1, 2$$

Inserting this into Eq. 3.C.5 for $i = 1$ and 2,

$$L - C - 0.5n \ln \sigma_1^2 \sigma_2^2 - 0.5 \frac{(n-1)(s_1^2 - 2s_{12} + s_2^2)}{\sigma_1^2 + \sigma_2^2} \qquad (3.C.9)$$

This is Eq. 3.3.4 in the text.

DERIVATION OF $L(\hat{\Omega})$, EQUATION 3.4.14

In Eqs. 3.A.5 and 3.A.10, replace σ_1^2, σ_2^2, and σ_μ^2 by their respective MLE's given in Eq. 3.3.3:

$$P = \frac{(n-1)^2\left(s_1^2 s_2^2 - s_{12}^2\right)}{n^2} \qquad (3.D.1)$$

$$L = C - 0.5n \ln P - \frac{0.5(n-1)^2 n^2 \left(s_1^2 s_2^2 - 2s_{12}^2 + s_1^2 s_2^2\right)}{(n-1)^2 n \left(s_1^2 s_2^2 - s_{12}^2\right)}$$

$$= C - 0.5n \ln P - n \qquad (3.D.2)$$

which is Eq. 3.4.14.

DERIVATION OF CONSTRAINED MAXIMUM LIKELIHOOD ESTIMATES IF ONE PRECISION ESTIMATE IS NEGATIVE

Reference is made to the development in Appendix 3.A with $\sigma_1^2 = 0$. The log likelihood function, Eq. 3.A.10, becomes

$$L = C - 0.5n \ln \sigma_\mu^2 \sigma_2^2 - \frac{0.5(n-1)\left[\left(\sigma_\mu^2 + \sigma_2^2\right)s_1^2 - 2\sigma_\mu^2 s_{12} + \sigma_\mu^2 s_2^2\right]}{\sigma_\mu^2 \sigma_2^2},$$

$$= C - 0.5n\left(\ln \sigma_\mu^2 + \ln \sigma_2^2\right) - 0.5(n-1)\left[\frac{s_1^2}{\sigma_\mu^2} + \frac{s_1^2 - 2s_{12} + s_2^2}{\sigma_2^2}\right]$$

$$\tag{3.E.1}$$

Then,

$$\frac{\partial L}{\partial \sigma_2^2} = \frac{-0.5n}{\sigma_2^2} + \frac{0.5(n-1)\left(s_1^2 - 2s_{12} + s_2^2\right)}{\sigma_2^4} = 0$$

from which

$$\hat{\sigma}_2^2 = \frac{(n-1)\left(s_1^2 - 2s_{12} + s_2^2\right)}{n} \tag{3.E.2}$$

which is Eq. 3.5.1. Also,

$$\frac{\partial L}{\partial \sigma_\mu^2} = \frac{-0.5n}{\sigma_\mu^2} + \frac{0.5(n-1)s_1^2}{\sigma_\mu^4} = 0$$

from which

$$\hat{\sigma}_\mu^2 = \frac{(n-1)s_1^2}{n} \tag{3.E.3}$$

which is Eq. 3.5.2.

DERIVATION OF CONSTRAINED MAXIMUM LIKELIHOOD ESTIMATES IF s_{12} IS NEGATIVE

Reference is made to the development in Appendix 3.A with $\sigma_\mu = 0$. The log likelihood function, Eq. 3.A.10, becomes

$$L = C - 0.5n \ln \sigma_1^2 \sigma_2^2 - \frac{0.5(n-1)\left(\sigma_2^2 s_1^2 + \sigma_1^2 s_2^2\right)}{\sigma_1^2 \sigma_2^2}$$

$$= C - 0.5n\left(\ln \sigma_1^2 + \ln \sigma_2^2\right) - 0.5(n-1)\left(\frac{s_1^2}{\sigma_1^2} + \frac{s_2^2}{\sigma_2^2}\right) \qquad (3.F.1)$$

Then, for $i = 1$ or 2,

$$\frac{\partial L}{\partial \sigma_i^2} = \frac{-0.5n}{\sigma_i^2} + \frac{0.5(n-1)s_i^2}{\sigma_i^4} = 0$$

from which

$$\hat{\sigma}_i^2 = \frac{(n-1)s_i^2}{n} \qquad (3.F.2)$$

which is Eq. 3.5.3.

DERIVATION OF CONSTRAINED EXPECTED LIKELIHOOD ESTIMATES RESULTS, SECTION 3.6

Equation 3.3.3 indicates that the MLE of $(\sigma_1^2 + \sigma_2^2)$ is S defined by Eq. 3.6.4. Denoting the CELE's by $\dot{\sigma}_1^2$ and $\dot{\sigma}_2^2$, respectively, they will by definition sum to S. Consider a point in the first quadrant on the line $(\sigma_1^2 + \sigma_2^2) = S$ and designate this point by

$$\sigma_{1v}^2 = vS; \qquad 0 \le v \le 1$$

$$\sigma_{2v}^2 = (1 - v)S \tag{3.G.1}$$

The CELE of σ_1^2 will be found by calculating L given by Eq. 3.A.10 as a function of v. Calling this L_v, the CELE of σ_1^2 is the weighted average

$$\dot{\sigma}_1^2 = \frac{S \int_0^1 v w_v \, dv}{\int_0^1 w_v \, dv} \tag{3.G.2}$$

where

$$w_v = \exp(L_v) \tag{3.G.3}$$

From Eq. 3.A.10,

$$L_v = C - 0.5n \ln P_v - \frac{0.5(n-1)A_v}{P_v} \tag{3.G.4}$$

where

$$A_v = \left(\sigma_{\mu v}^2 + \sigma_{2v}^2\right)s_1^2 - 2\sigma_{\mu v}^2 s_{12} + \left(\sigma_{\mu v}^2 + \sigma_{1v}^2\right)s_2^2 \tag{3.G.5}$$

$$P_v = S\left(\sigma_{\mu v}^2 + vS - v^2S\right) \tag{3.G.6}$$

and where $\sigma_{\mu v}^2$ is calculated from Eq. 3.3.2 with σ_1^2 and σ_2^2 replaced by σ_{1v}^2 and σ_{2v}^2, respectively. The expression for $\sigma_{\mu v}^2$ is

$$\sigma_{\mu v}^2 = vS(2v - 1) + \frac{(n - 1)\left[2v\left(s_{12} - s_1^2\right) + s_1^2\right]}{n} \qquad (3.G.7)$$

Inserting this expression in Eqs. 3.G.5 and 3.G.6 and replacing S by $(n - 1)(s_1^2 + s_2^2 - 2s_{12})/n$ as given by Eq. 3.6.2, we have

$$\frac{0.5(n - 1)A_v}{P_v} = n \qquad (3.G.8)$$

Hence,

$$w_v = e^c e^{-n} e^{-0.5n \ln P_v}$$

$$= c_1 P_v^{-0.5n} \qquad (3.G.9)$$

and from Eq. 3.G.2,

$$\dot{\sigma}_1^2 = \frac{S\int_0^1 vP_v^{-0.5n}\,dv}{\int_0^1 P_v^{-0.5n}\,dv} \qquad (3.G.10)$$

where P_v reduces to

$$P_v = \frac{n - 1}{n}S\left[s_1^2(1 - v)^2 + v^2 s_2^2 + 2v(1 - v)s_{12}\right] \qquad (3.G.11)$$

The factor $(n - 1)S/n$ appears in both the numerator and denominator of Eq. 3.G.10. The expression in brackets is rewritten as a quadratic function of v:

$$g(v) = \left(s_1^2 + s_2^2 - 2s_{12}\right)v^2 + 2\left(s_{12} - s_1^2\right)v + s_1^2$$

$$= \frac{nSv^2}{n - 1} + 2\left(s_{12} - s_1^2\right)v + s_1^2 \qquad (3.G.12)$$

Writing

$$f(v) = [g(v)]^{-0.5n} \qquad (3.G.13)$$

the expression for $\dot{\sigma}_1^2$ is

$$\dot{\sigma}_1^2 = \frac{S\int_0^1 vf(v)\,dv}{\int_0^1 f(v)\,dv} \qquad (3.G.14)$$

which is Eq. 3.6.1 in the text, where $f(v)$ is defined by Eq. 3.G.13 or by Eq. 3.6.5 in the text.

REFERENCES

1. Grubbs, F. E. "On Estimating Precision of Measuring Instruments and Product Variability." *J. Amer. Stat. Assn.* **43**: 243–264; 1948.

2. Thompson, W. A., Jr. "Precision of Simultaneous Measurement Procedures." *J. Amer. Stat. Assn.* **58**: 474–479; 1963.

3. Jaech, J. L. "Large Sample Tests of Instrument Precision: More Than Two Instruments." *Technometrics.* **18**(2): 127–133; 1976.

4. Dowdy, E. J., Nicholson, N., and Caldwell, J. T. "A Nonintrusive Irradiated Fuel Inventory Confirmation Technique." *ESARDA Proceedings: Second Annual Symposium on Safeguards and Nuclear Material Management.* 1980; pp. 353–358.

5. Maloney, C. J. and Rastogi, S. C. "Significance Test for Grubbs's (sic) Estimators." *Biometrics.* **26**: 671–676; 1970.

6. Bennett, C. A. and Franklin, N. L. *Statistical Analysis in Chemistry and the Chemical Industry.* New York: John Wiley and Sons; 1954; p. 132.

7. Jaech, J. L. "Further Tests of Significance for Grubbs' Estimators." *Biometrics.* **27**: 1097–1101; 1971.

8. Thompson, W. A., Jr. "The Problem of Negative Estimates of Variance Components." *Ann. Math. Stat.* **33**: 273–89; 1962.

9. Jaech, J. L. "Constrained Expected Likelihood Estimates of Precisions Using Grubbs' Technique for Two Measurement Methods." *Nucl. Mat. Manage. J.* **X**(2): 34–39; 1981.

Chapter 4

$N = 3$ MEASUREMENT METHODS: INFERENCE BASED ON PAIRED DIFFERENCES

4.1 INTRODUCTION

With reference to the model given as Eq. 1.8.1, in this chapter we study the case of $N = 3$ measurement methods with constant bias and different precisions: $\beta_1 = \beta_2 = \beta_3$, $\alpha_1 \neq \alpha_2 \neq \alpha_3$, and $\sigma_1^2 \neq \sigma_2^2 \neq \sigma_3^2$. The model becomes

$$x_{ik} = \alpha_i + \mu_k + \varepsilon_{ik} \tag{4.1.1}$$

In Chapters 4 and 5, we operate on the paired differences:

$$y_{ijk} = x_{ik} - x_{jk}$$

$$= (\alpha_i - \alpha_j) + (\varepsilon_{ik} - \varepsilon_{jk}) \tag{4.1.2}$$

Note from Eq. 4.1.2 that μ_k does not appear, and hence a distinction need not be made between the random and fixed models. Inference based on the x_{ik}'s rather than the y_{ijk}'s is presented in Chapter 6, at which point the advantages and disadvantages of both approaches are discussed.

In Chapter 4, attention is restricted to $N = 3$, with the discussion on $N \geq 4$ deferred to Chapter 5. Although there is a commonality between results such that the two chapters could logically be combined, they are kept separate for a number of reasons: (1) the derivations are easier to follow for $N = 3$ than for general N, and such derivations in Chapter 4 lead to a simpler exposition of the material for $N \geq 4$ in Chapter 5; (2) there is a relationship between Grubbs' estimators and the maximum likelihood estimators (MLE's) for $N = 3$ that does not exist for $N \geq 4$; (3) less importantly, from the viewpoint of applications, problems involving $N = 3$ are more common than those for $N \geq 4$, judging from my experience, and it is worthwhile to address this model specifically. In Chapter 6, dealing with the original x_{ik} values, discussions for $N = 3$ and $N \geq 4$ are combined.

4.2 GRUBBS' ESTIMATORS

As mentioned in the discussion of $N = 2$, Grubbs' estimators are based on the method of moments, in which sample moments are equated to population moments with the resulting equations solved in some sense. With y_{ijk} given by Eq. 4.1.2, define

$$V_{ij} = \frac{\sum\limits_{k=1}^{n} y_{ijk}^2 - \left(\sum\limits_{k=1}^{n} y_{ijk}\right)^2 \Big/ n}{n - 1} \; ; \qquad i = 1, 2; \qquad j > i \qquad (4.2.1)$$

It follows from Eq. 4.1.2 that

$$E(V_{ij}) = \sigma_i^2 + \sigma_j^2 \qquad (4.2.2)$$

For $N = 3$, there are three combinations of i and j, and hence the three moment equations to be solved are

$$V_{12} = \sigma_1^2 + \sigma_2^2$$

$$V_{13} = \sigma_1^2 + \sigma_3^2 \qquad (4.2.3)$$

$$V_{23} = \sigma_2^2 + \sigma_3^2$$

The solutions are easily found and result in the Grubbs' (1) estimators based on differences:

$$\tilde{\sigma}_1^2 = 0.5(V_{12} + V_{13} - V_{23})$$

$$\tilde{\sigma}_2^2 = 0.5(V_{12} - V_{13} + V_{23}) \qquad (4.2.4)$$

$$\tilde{\sigma}_3^2 = 0.5(-V_{12} + V_{13} + V_{23})$$

Before these results are illustrated, the MLE's are discussed.

4.3 MAXIMUM LIKELIHOOD ESTIMATORS (MLE'S)

The MLE of σ_1^2 is derived in Appendix 4.A, and is

$$\hat{\sigma}_1^2 = \frac{(n-1)\left[(\sigma_2^2 + \sigma_3^2)(\sigma_3^2 V_{12} + \sigma_2^2 V_{13}) - \sigma_2^2 \sigma_3^2 V_{23}\right]}{n(\sigma_2^2 + \sigma_3^2)^2} - \frac{\sigma_2^2 \sigma_3^2}{\sigma_2^2 + \sigma_3^2}$$

$$(4.3.1)$$

The MLE's of σ_2^2 and σ_3^2 follow by analogy. Note that $\hat{\sigma}_1^2$ is a function of σ_2^2 and σ_3^2, and the three equations in the three unknowns appear formidable to solve. It is shown in Appendix 4.B, however, that

$$\hat{\sigma}_i^2 = \frac{(n-1)\tilde{\sigma}_i^2}{n}; \qquad i = 1, 2, 3 \qquad (4.3.2)$$

In other words, the MLE's are $(n-1)/n$ times the Grubbs' estimators. It will be recalled that the same relationship between Grubbs' estimators and the MLE's held for the random model for $N = 2$.

EXAMPLE 4.A

In Example 3.A, measurement data were given for 2 methods of measuring percent theoretical densities (%TD) of 43 sintered uranium fuel pellets. From this example, the Grubbs' estimates of the parameters σ_1^2 and σ_6^2 were 0.010016 and 0.034486 (%TD)^2 for σ_1^2 and σ_6^2, respectively. The data for a third measurement method, identified as Method 3 in the cited reference (Reference 3 in Chapter 3) are given in Table 4.1 in the same pellet order as in Table 3.1.

After finding the three columns of differences, y_{ijk}, applying Eq. 4.2.1 gives

$$V_{16} = 0.044502 \qquad V_{13} = 0.030374 \qquad V_{63} = 0.066511$$

From Eq. 4.2.4, the Grubbs' estimates are

$$\tilde{\sigma}_1^2 = 0.5(0.044502 + 0.030374 - 0.066511) = 0.004183$$

$$\tilde{\sigma}_6^2 = 0.040320 \qquad \tilde{\sigma}_3^2 = 0.026192$$

Table 4.1   Pellet Densities for Method 3

4.14	4.70	4.07	4.64
4.53	3.85	4.19	4.83
4.57	4.13	4.33	4.60
4.75	3.94	4.36	4.66
4.84	4.50	4.57	4.07
4.21	4.98	4.61	4.54
4.01	4.33	4.22	4.35
4.47	4.47	4.69	4.68
4.09	4.57	4.45	4.45
4.60	4.58	4.44	4.60
4.40	4.45	4.28	

and from Eq. 4.3.2, the MLE's are 42/43 times these respective values:

$$\hat{\sigma}_1^2 = 0.004086 \qquad \hat{\sigma}_6^2 = 0.039382 \qquad \hat{\sigma}_3^2 = 0.025583 \qquad \blacksquare$$

EXAMPLE **4.B**

In Example 3.B, which deals with the image extinction method of measuring Cerenkov glow intensity, the data for Observers 2 and 3 were given. The Grubbs' estimates of the parameters were 659.57×10^{-10} and 441.59×10^{-10} for σ_2^2 and σ_3^2, respectively. The data for a third observer, identified as Observer 1 in the cited reference (Reference 4 in Chapter 3) are given in Table 4.2 in the same order as in Table 3.2.

After finding the three columns of differences, y_{ijk}, applying Eq. 4.2.1 gives

$$V_{12} = 1076.55 \times 10^{-10} \qquad V_{13} = 1019.99 \times 10^{-10}$$

$$V_{23} = 1101.16 \times 10^{-10}$$

From Eq. 4.2.4, the Grubbs' estimates are

$$\tilde{\sigma}_1^2 = 0.5(1076.55 + 1019.99 - 1101.16) \times 10^{-10} = 497.69 \times 10^{-10}$$

$$\tilde{\sigma}_2^2 = 578.86 \times 10^{-10} \qquad \tilde{\sigma}_3^2 = 522.30 \times 10^{-10}$$

and from Eq. 4.3.2, the MLE's are 35/36 times these respective values:

$$\hat{\sigma}_1^2 = 483.87 \times 10^{-10} \qquad \hat{\sigma}_2^2 = 562.78 \times 10^{-10} \qquad \hat{\sigma}_3^2 = 507.79 \times 10^{-10} \qquad \blacksquare$$

Table 4.2   Burnup/Aperture Size for Observer 1

1.64×10^{-3}	2.18	1.90
1.65	2.59	2.10
1.65	1.49	1.85
1.18	1.93	2.04
1.61	1.89	1.55
1.57	1.90	1.13
1.68	2.08	1.95
1.98	1.89	2.15
1.99	1.98	1.48
2.18	2.31	1.49
2.52	2.06	1.91
2.07	2.13	1.75

Table 4.3   Fuse Burning Times (secs) for Observer C (3)

10.07	9.64	9.55
9.90	10.24	9.54
9.86	9.86	9.88
9.70	9.63	9.51
9.65	9.65	9.53
9.83	9.74	9.45
9.79	10.34	9.67
9.59	9.86	9.78
9.72	9.65	9.86
9.92	9.50	

EXAMPLE 4.C

We return to the fuse burning example reported by Grubbs (1) for which data from Observers A and B (1 and 2) were given in Table 3.3 of Example 3.M. The data for the third observer, C (3), are given in Table 4.3. Recall from Example 3.M that the estimate of σ_2^2 was negative when the data for only the first two observers were used. The data ordering is the same as in Table 3.3.

After finding the three columns of differences, y_{ijk}, applying Eq. 4.2.1 gives

$$V_{12} = 0.000703 \qquad V_{13} = 0.000889 \qquad V_{23} = 0.000311$$

From Eq. 4.2.4, the Grubbs' estimates are

$$\tilde{\sigma}_1^2 = 0.5(0.000703 + 0.000889 - 0.000311) = 0.000641$$

$$\tilde{\sigma}_2^2 = 0.000063 \qquad \tilde{\sigma}_3^2 = 0.000249$$

The corresponding MLE's are $28/29$ times these values, from Eq. 4.3.2:

$$\hat{\sigma}_1^2 = 0.000619 \qquad \hat{\sigma}_2^2 = 0.000061 \qquad \hat{\sigma}_3^2 = 0.000240 \qquad \blacksquare$$

EXAMPLE 4.D

Mandel (2) gives data from an interlaboratory study measuring the stress at 600% elongation of seven rubber samples ($n = 7$). Thirteen laboratories participated in the study ($N = 13$), and each made 4 replicate measurements on each material. In this example, we restrict attention to Laboratories 6, 8, and 12, for reasons that are made clear in later chapters. Also, the data are transformed to natural logarithms to stabilize the variance, and the measured

Table 4.4   Logarithms of Stress of Rubber Samples (kg/cm²)

Sample	Lab 6	Lab 8	Lab 12
1	3.906	4.196	4.278
2	4.718	4.960	4.873
3	3.400	3.628	3.502
4	3.917	4.162	4.142
5	3.239	3.520	3.402
6	3.228	3.487	3.409
7	3.700	3.848	3.812

value used in the analysis is the average of the four logarithms per cell. These averages are given by Jaech (3). The data are given in Table 4.4. After calculating the difference y_{ijk}, applying Eq. 4.2.1 gives

$$V_{68} = 0.001742 \qquad V_{6,12} = 0.005448 \qquad V_{8,12} = 0.005153$$

From Eq. 4.2.4, the Grubbs' estimates are

$$\tilde{\sigma}_6^2 = 0.001019 \qquad \tilde{\sigma}_8^2 = 0.000724 \qquad \tilde{\sigma}_{12}^2 = 0.004430$$

and, from Eq. 4.3.2, the MLE's are 6/7 times these respective estimates:

$$\hat{\sigma}_6^2 = 0.000873 \qquad \hat{\sigma}_8^2 = 0.000621 \qquad \hat{\sigma}_{12}^2 = 0.003797$$

Note: These are the estimated measurement variances on the logarithmically transformed scale. To convert them to the original scale (kg/cm²), use is made of the following well known result (4): If $y = \ln x$ is normally distributed with mean μ and variance σ^2, then the mean of x is $\exp(\mu + \sigma^2/2)$, and the variance of x is $\exp[2\mu + \sigma^2(\exp\sigma^2 - 1)]$. Thus on a *relative* basis, the standard deviation of x is $(\exp\sigma^2 - 1)^{0.5}$. Considering the MLE of Lab 12, for example, for which σ_{12}^2 is estimated to be 0.003797, the relative standard deviation of the measured stress is $(\exp 0.003797 - 1)^{0.5} = 0.062$, or 6.2% relative. Thus at a measured stress of 30 kg/cm², the estimated standard deviation is 1.86 kg/cm², while at a stress of 120 kg/cm², it is 7.44 kg/cm². ∎

These examples are reconsidered later from different perspectives.

4.4   INFERENCE

Statistical inference problems include both constructing interval estimates and testing hypotheses.

4.4.1 Interval Estimation

Confidence Interval on σ_i^2. In his article, Grubbs (1) gives an expression for the variance of $\tilde{\sigma}_i^2$ of Eq. 4.2.4, for $i = 1, 2, 3$:

$$\text{var } \tilde{\sigma}_i^2 = \frac{2\sigma_i^4}{n-1} + \frac{\sigma_1^2\sigma_2^2 + \sigma_1^2\sigma_3^2 + \sigma_2^2\sigma_3^2}{n-1} \tag{4.4.1}$$

Comparing this result with the variance of $\tilde{\sigma}_i^2$ for $N = 2$, given by Eq. 3.4.1, shows that the expressions are identical except that σ_μ^2 of Eq. 3.4.1 is replaced by σ_3^2. Thus the estimate of σ_i^2 will have smaller variance for $N = 3$ than it will for $N = 2$ if σ_3^2, the measurement error variance for the third method, is smaller than σ_μ^2, the process variance.

Calculating $\text{var } \tilde{\sigma}_i^2$ from Eq. 4.4.1 requires knowing the values of the unknown parameters. They may be replaced by their estimates to obtain an estimate of $\text{var } \tilde{\sigma}_i^2$. With $\text{var } \tilde{\sigma}_i^2$ given by Eq. 4.4.1, Eqs. 3.4.2 and 3.4.3 may be applied to construct an approximate $100(p_2 - p_1)\%$ confidence interval about σ_i^2.

EXAMPLE 4.E

In Example 4.A, Grubbs' estimates of the parameters were found to be

$$\tilde{\sigma}_1^2 = 0.004183 \qquad \tilde{\sigma}_3^2 = 0.026192 \qquad \tilde{\sigma}_6^2 = 0.040320$$

From Eq. 4.4.1, their respective sampling variances are estimated by

$$\text{var } \tilde{\sigma}_1^2 = \{2(0.4183)^2 + [(0.4183)(2.6192) + (0.4183)(4.0320)$$

$$+ (2.6192)(4.0320)]\} \times 10^{-4}/42$$

$$= 0.3260 \times 10^{-4}$$

$$\text{var } \tilde{\sigma}_3^2 = 0.6444 \times 10^{-4} \qquad \text{var } \tilde{\sigma}_6^2 = 1.0918 \times 10^{-4}$$

The approximate degrees of freedom are, from Eq. 3.4.2,

$$df_1 = 1.1 \qquad df_3 = 21.3 \qquad df_6 = 29.8$$

When the degrees of freedom are fewer than, say, 2, it is prudent to simply remark that the parameter in question has been poorly estimated. Formal construction of the confidence limits would lead to a very broad interval that should probably be interpreted only qualitatively. In constructing 95% confidence limits for σ_3^2 and σ_6^2, from Table 1.2:

$$\chi_{0.025}^2(21.3) = 10.5 \qquad \chi_{0.975}^2(21.3) = 35.9$$

$$\chi_{0.025}^2(29.8) = 16.6 \qquad \chi_{0.975}^2(29.8) = 46.7$$

From Eq. 3.4.3, the approximate 95% confidence limits are

$$\frac{(21.3)(0.026192)}{35.9} < \sigma_3^2 < \frac{(21.3)(0.026192)}{10.5}$$

$$0.0155 < \sigma_3^2 < 0.0531$$

$$0.0257 < \sigma_6^2 < 0.0724 \qquad\qquad \blacksquare$$

EXAMPLE 4.F

In Example 4.B, Grubbs' estimates of the parameters were

$$\tilde{\sigma}_1^2 = 497.69 \times 10^{-10} \qquad \tilde{\sigma}_2^2 = 578.86 \times 10^{-10} \qquad \tilde{\sigma}_3^2 = 522.30 \times 10^{-10}$$

From Eq. 4.4.1, their respective sampling variances are estimated by

$$\operatorname{var} \tilde{\sigma}_1^2 = \frac{1}{35} \{ 2(497.69)^2 + [(497.69)(578.86) + (497.69)(522.30)$$
$$+ (578.86)(522.30)] \} \times 10^{-20}$$
$$= 38,450 \times 10^{-20}$$
$$\operatorname{var} \tilde{\sigma}_2^2 = 43,444 \times 10^{-20} \qquad \operatorname{var} \tilde{\sigma}_3^2 = 39,885 \times 10^{-20}$$

From Eq. 3.4.2, the approximate degrees of freedom are

$$df_1 = 12.9 \qquad df_2 = 15.4 \qquad df_3 = 13.7$$

In constructing approximate 95% confidence limits, from Table 1.2,

$$\chi_{0.025}^2(12.9) = 4.95 \qquad \chi_{0.975}^2(12.9) = 24.60$$

$$\chi_{0.025}^2(15.4) = 6.52 \qquad \chi_{0.975}^2(15.4) = 28.03$$

$$\chi_{0.025}^2(13.7) = 5.44 \qquad \chi_{0.975}^2(13.7) = 25.71$$

From Eq. 3.4.3, the approximate 95% confidence limits are

$$\frac{(12.9)(497.69) \times 10^{-10}}{24.60} < \sigma_1^2 < \frac{(12.9)(497.69) \times 10^{-10}}{4.95}$$

$$261 \times 10^{-10} < \sigma_1^2 < 1297 \times 10^{-10}$$

$$318 \times 10^{-10} < \sigma_2^2 < 1367 \times 10^{-10}$$

$$278 \times 10^{-10} < \sigma_3^2 < 1315 \times 10^{-10}$$ ■

EXAMPLE 4.G

Approximate 95% confidence limits are constructed on the parameters for Examples 4.C and 4.D. Some intermediate calculations and the results follow. The computational steps are identical to those in Examples 4.E and 4.F.
   From Example 4.C,

$$n = 29 \qquad \tilde{\sigma}_1^2 = 0.000641 \qquad \tilde{\sigma}_2^2 = 0.000063 \qquad \tilde{\sigma}_3^2 = 0.000249$$

$$\text{var } \tilde{\sigma}_1^2 = 3.71 \times 10^{-8} \qquad \text{var } \tilde{\sigma}_2^2 = 0.80 \times 10^{-8} \qquad \text{var } \tilde{\sigma}_3^2 = 1.21 \times 10^{-8}$$

$$df_1 = 22.1 \qquad df_2 = 1.0 \qquad df_3 = 10.2$$

It is apparent that σ_2^2 has been poorly estimated. For σ_1^2 and σ_3^2,

$$\chi_{0.025}^2(22.1) = 11.1 \qquad \chi_{0.975}^2(22.1) = 36.9$$

$$\chi_{0.025}^2(10.2) = 3.36 \qquad \chi_{0.975}^2(10.2) = 20.77$$

$$0.000384 < \sigma_1^2 < 0.001276$$

$$0.000122 < \sigma_3^2 < 0.000756$$

From Example 4.D,

$$n = 7 \quad \tilde{\sigma}_6^2 = 0.001019 \quad \tilde{\sigma}_8^2 = 0.000724 \quad \tilde{\sigma}_{12}^2 = 0.004430$$

$$\text{var } \tilde{\sigma}_6^2 = 175.60 \times 10^{-8} \quad \text{var } \tilde{\sigma}_8^2 = 158.46 \times 10^{-8} \quad \text{var } \tilde{\sigma}_{12}^2 = 795.15 \times 10^{-8}$$

$$df_6 = 1.2 \quad df_8 = 0.7 \quad df_{12} = 4.9$$

No attempt is made to construct confidence limits on σ_6^2 or σ_8^2. For σ_{12}^2,

$$\chi_{0.025}^2(4.9) = 0.80 \qquad \chi_{0.975}^2(4.9) = 11.5$$

$$0.001888 < \sigma_{12}^2 < 0.027134 \qquad\qquad \blacksquare$$

Confidence Intervals Based on Large-Sample Theory. Large-sample theory may be used to find the asymptotic variance of $\hat{\sigma}_i^2$ as well as to construct confidence regions on all of the parameters. Since the Grubbs' estimators and the MLE's are in agreement within the $(n - 1)/n$ factor, we would expect good agreement between the asymptotic variance of $\hat{\sigma}_i^2$ based on large-sample theory and the variance of $\tilde{\sigma}_i^2$ given by Eq. 4.4.1.

By large-sample theory (5), the quantity u, defined in Eq. 4.4.2, is distributed approximately as chi-square with 3 degrees of freedom:

$$u = a_{11}(\hat{\sigma}_1^2 - \sigma_1^2)^2 + a_{22}(\hat{\sigma}_2^2 - \sigma_2^2)^2$$

$$+ a_{33}(\hat{\sigma}_3^2 - \sigma_3^2)^2 + 2a_{12}(\hat{\sigma}_1^2 - \sigma_1^2)(\hat{\sigma}_2^2 - \sigma_2^2)$$

$$+ 2a_{13}(\hat{\sigma}_1^2 - \sigma_1^2)(\hat{\sigma}_3^2 - \sigma_3^2) + 2a_{23}(\hat{\sigma}_2^2 - \sigma_2^2)(\hat{\sigma}_3^2 - \sigma_3^2) \quad (4.4.2)$$

The quantity a_{ij} is the element in the ith row and jth column of the matrix:

$$H = \frac{0.5n}{Q^2} \begin{vmatrix} (\hat{\sigma}_2^2 + \hat{\sigma}_3^2)^2 & \hat{\sigma}_3^4 & \hat{\sigma}_2^4 \\ \hat{\sigma}_3^4 & (\hat{\sigma}_1^2 + \hat{\sigma}_3^2)^2 & \hat{\sigma}_1^4 \\ \hat{\sigma}_2^4 & \hat{\sigma}_1^4 & (\hat{\sigma}_1^2 + \hat{\sigma}_2^2)^2 \end{vmatrix} \quad (4.4.3)$$

where Q is defined by Eq. 4.A.5 with $\hat{\sigma}_i^2$ replacing σ_i^2.

The variance of $\hat{\sigma}_i^2$ is the element in position (i, i) of H^{-1}. These results are derived in Appendix 4.C. Their application is illustrated in the following example.

EXAMPLE 4.H

With reference to Example 4.A, the variances of the Grubbs' estimators were found in Example 4.E to be

$$\text{var } \tilde{\sigma}_1^2 = 0.3260 \times 10^{-4} \qquad \text{var } \tilde{\sigma}_3^2 = 0.6444 \times 10^{-4}$$

$$\text{var } \tilde{\sigma}_6^2 = 1.0918 \times 10^{-4}$$

For these data,

$$n = 43 \qquad \hat{\sigma}_1^2 = 0.004086 \qquad \hat{\sigma}_3^2 = 0.025583 \qquad \hat{\sigma}_6^2 = 0.039382$$

$$Q = 0.001273$$

Then, by Eq. 4.4.3,

$$H = 13,268,175 \begin{pmatrix} 0.00422045 & 0.00155094 & 0.00065449 \\ 0.00155094 & 0.00188947 & 0.00001670 \\ 0.00065449 & 0.00001670 & 0.00088025 \end{pmatrix}$$

The inverse matrix is

$$H^{-1} = 10^{-4} \begin{pmatrix} 0.3038 & -0.2474 & -0.2212 \\ -0.2474 & 0.6004 & 0.1726 \\ -0.2212 & 0.1726 & 1.0174 \end{pmatrix}$$

Note how the diagonal elements compare with $\text{var}\,\tilde{\sigma}_1^2$, $\text{var}\,\tilde{\sigma}_3^2$, and $\text{var}\,\tilde{\sigma}_6^2$, respectively, as given above.

An approximate 90% region may be constructed on $(\sigma_1^2, \sigma_3^2, \sigma_6^2)$ by finding those combinations of parameter values for which the quantity u in Eq. 4.4.2 is less than 6.25, where, from Table 1.2,

$$\Pr[\chi^2(3) < 6.25] = 0.90$$

To illustrate, with the appropriate subscript changes,

$$u = 55,998(\hat{\sigma}_1^2 - \sigma_1^2)^2 + 25,070(\hat{\sigma}_3^2 - \sigma_3^2)^2 + 41,156(\hat{\sigma}_1^2 - \sigma_1^2)(\hat{\sigma}_3^2 - \sigma_3^2)$$

$$+ \text{terms in}(\hat{\sigma}_6^2 - \sigma_6^2)$$

Find the limits on σ_1^2 for $\sigma_6^2 = \hat{\sigma}_6^2$ and for three cases:

$$\sigma_3^2 = \hat{\sigma}_3^2; \qquad \sigma_3^2 = 0.040; \qquad \sigma_3^2 = 0.015$$

For $\sigma_3^2 = \hat{\sigma}_3^2$,

$$u = 55,998(\hat{\sigma}_1^2 - \sigma_1^2)^2 < 6.25$$

$$|\hat{\sigma}_1^2 - \sigma_1^2| < 0.01056$$

$$0 < \sigma_1^2 < 0.01465$$

For $\sigma_3^2 = 0.040$

$$u = 55{,}998\left(\hat{\sigma}_1^2 - \sigma_1^2\right)^2 + 5.21 - 593.35\left(\hat{\sigma}_1^2 - \sigma_1^2\right) < 6.25$$

$$\hat{\sigma}_1^2 - \sigma_1^2 = -0.00153$$

and

$$\hat{\sigma}_1^2 - \sigma_1^2 = 0.01213$$

$$0 < \sigma_1^2 < 0.00562$$

For $\sigma_3^2 = 0.015$,

$$u = 55{,}998\left(\hat{\sigma}_1^2 - \sigma_1^2\right)^2 + 2.81 + 435.55\left(\hat{\sigma}_1^2 - \sigma_1^2\right) < 6.25$$

$$\hat{\sigma}_1^2 - \sigma_1^2 = -0.01173$$

and

$$\hat{\sigma}_1^2 - \sigma_1^2 = 0.00395$$

$$0.00014 < \sigma_1^2 < 0.01582$$

The three-dimensional confidence region on $(\sigma_1^2, \sigma_3^2, \sigma_6^2)$ can be found by proceeding as indicated in the sample calculations above. ∎

4.4.2 Hypothesis Testing

The hypotheses tests discussed here are large-sample likelihood ratio tests formulated by Jaech (6) for $N \geq 3$ methods. For the specific case $N = 3$, the likelihood ratio test was considered earlier by Russell and Bradley (7).

The logarithm of the likelihood, denoted by L, is given in Appendix 4.A as Eq. 4.A.12, with Q defined in Eq. 4.A.5. To test any given hypothesis about the σ_i's, the maximum value of L is found under that hypothesis and denoted by $L(\hat{\omega})$. The value of L corresponding to the MLE's of the σ_i^2's is calculated and denoted by $L(\hat{\Omega})$. Not including the α_i's that are estimated in both spaces, the number of parameters estimated in Ω space is 3. Letting r be the number of parameters estimated in ω space, large-sample distribution theory tells us that

$$\lambda = -2\left[L(\hat{\omega}) - L(\hat{\Omega})\right] \tag{4.4.4}$$

is distributed approximately as chi-square with $(3 - r)$ degrees of freedom.

In Appendix 4.D, it is shown that

$$L(\hat{\Omega}) = C - 0.5n \ln Q - n \qquad (4.4.5)$$

where Q is defined by Eq. 4.A.5 with the σ_i^2's replaced by their MLE's given by Eqs. 4.2.4 and 4.3.2.

The calculation of $L(\hat{\omega})$ depends on the null hypothesis being tested. One hypothesis of particular interest is that the three measurement precisions are equal:

$$H_{01}: \quad \sigma_1^2 = \sigma_2^2 = \sigma_3^2 \ (= \sigma_0^2)$$

Under H_{01}, Appendix 4.E shows that the MLE of σ_0^2 is

$$\hat{\sigma}_0^2 = \frac{(n-1)(V_{12} + V_{13} + V_{23})}{6n} \qquad (4.4.6)$$

and that the likelihood is

$$L(\hat{\omega}) = C - 0.5n\left(\ln 3 + 2 \ln \hat{\sigma}_0^2\right) - n \qquad (4.4.7)$$

Other hypotheses are considered in the examples to follow.

EXAMPLE 4.1

In Example 4.C, which deals with the measurement of fuse burning times by three observers, the MLE's of σ_1^2, σ_2^2, and σ_3^2 were 0.000619 sec^2, 0.000061 sec^2, and 0.000240 sec^2, respectively. Test the hypothesis at $\alpha = 0.05$ significance that $\sigma_1^2 = \sigma_2^2 = \sigma_3^2 \ (= \sigma_0^2)$. First, $L(\hat{\Omega})$ is calculated by Eq. 4.4.5, where, by Eq. 4.A.5,

$$Q = 10^{-8}[(6.19)(0.61) + (6.19)(2.40) + (0.61)(2.40)]$$

$$= 20.0959 \times 10^{-8}$$

For $n = 29$,

$$L(\hat{\Omega}) = C - 14.5 \ln(20.0959) \times 10^{-8} - 29$$

$$= C + 194.592$$

Then, $L(\hat{\omega})$ is calculated from Eq. 4.4.7 with $\hat{\sigma}_0^2$ calculated from Eq. 4.4.6:

$$\hat{\sigma}_0^2 = 3.07 \times 10^{-4}$$

(note that $\hat{\sigma}_0^2$ is the mean of $\hat{\sigma}_1^2$, $\hat{\sigma}_2^2$, and $\hat{\sigma}_3^2$, the MLE's)

$$L(\hat{\omega}) = C - 14.5(\ln 3 + 2\ln 3.07 \times 10^{-4}) - 29$$

$$= C + 189.641$$

From Eq. 4.4.4,

$$\lambda = -2(189.641 - 194.592) = 9.90$$

Since one parameter is estimated under the hypothesis, λ is distributed approximately as chi-square with 2 degrees of freedom. For $\alpha = 0.05$, the critical value read from Table 1.2 is 5.99. Since 9.90 > 5.99, the hypothesis is rejected; the evidence suggests that the observers have different measurement precisions.   ∎

EXAMPLE 4.J

In Example 4.A dealing with the measurement of densities of sintered uranium fuel pellets, Methods 1 and 3 were geometric methods while Method 6 was an immersion method. For $\alpha = 0.05$, test the hypothesis

$$H_{02}: \quad \sigma_1^2 = \sigma_3^2 \; (= \sigma_0^2)$$

for $\alpha = 0.05$ significance level.

Recall that $n = 43$ and that the respective MLE's for σ_1^2, σ_3^2, and σ_6^2 were 0.004086, 0.025583, and 0.039382.

It can be shown that the MLE of σ_0^2 is the mean of the MLE's of σ_1^2 and σ_3^2:

$$\hat{\sigma}_0^2 = 0.014835$$

and the MLE of σ_6^2 remains 0.039382.

First, $L(\hat{\Omega})$ is calculated from Eq. 4.4.5 where, by Eq. 4.A.5,

$$Q = [(40.86)(255.83) + \cdots + (255.83)(393.82)] \times 10^{-8} = 127{,}296 \times 10^{-8}$$

Then,

$$L(\hat{\Omega}) = C - 21.5\ln(127{,}296 \times 10^{-8}) - 43$$

$$= C + 100.328$$

Next, $L(\hat{\omega})$ is calculated from Eq. 4.A.12 with $\sigma_1^2 = \sigma_3^2 = 0.014835$ and $\sigma_6^2 = 0.039382$. First, again using Eq. 4.A.5,

$$Q = \left[2(148.35)(393.82) + (148.35)^2 \right] \times 10^{-8} = 138{,}854 \times 10^{-8}$$

As was the case under H_{01}, the last term in $L(\hat{\omega})$ is again $-n$ so that

$$L(\hat{\omega}) = C - 21.5 \ln(138,854 \times 10^{-8}) - 43$$

$$= C + 98.459$$

From Eq. 4.4.4,

$$\lambda = -2(98.459 - 100.328) = 3.74$$

There is $(3 - 2) = 1$ degree of freedom. From Table 1.2, the $\alpha = 0.05$ critical value is 3.84. Since λ is smaller than the critical value, H_{02} is not rejected; there is no reason to conclude that the two geometric methods have different precisions. ∎

EXAMPLE 4.K

In the preceding example, test the hypothesis

$$H_{03}: \quad \sigma_1^2 = 0.01, \qquad \sigma_3^2 = 0.01, \qquad \sigma_6^2 = 0.06$$

against the alternative that at least one of the equality signs is not valid. Use $\alpha = 0.05$.

The quantity $L(\hat{\Omega})$ was calculated in Example 4.J:

$$L(\hat{\Omega}) = C + 100.328$$

Under H_{03}, $L(\hat{\omega})$ is calculated by evaluating Eq. 4.A.12 for

$$\sigma_1^2 = 0.01, \qquad \sigma_3^2 = 0.01, \qquad \sigma_3^2 = 0.06$$

From Eq. 4.A.5,

$$Q = [(1)(1) + (1)(6) + (1)(6)] \times 10^{-4} = 13 \times 10^{-4}$$

Then

$$L(\hat{\omega}) = C - 21.5 \ln Q - \frac{21[(6)(3.0374) + (4.4502) + (6.6511)]}{13}$$

$$= C + 95.504$$

From Eq. 4.4.4

$$\lambda = -2(95.504 - 100.328) = 9.65$$

At $\alpha = 0.05$, with $(3 - 0) = 3$ degrees of freedom, the critical value from Table 1.2 is 7.81. Since $\lambda > 7.81$, the hypothesis H_{03} is rejected; at least one of the equality signs is not valid.   ■

4.5 CONSTRAINED MAXIMUM LIKELIHOOD ESTIMATION (CMLE)

One of the estimates of σ_i^2 may be negative. It is not possible for more than one estimate to be negative, as is proven in Appendix 4.F. If $\hat{\sigma}_i^2$ is negative, the estimate of σ_i^2 becomes zero. The CMLE of σ_j^2 for $j \neq i$ is then derived in Appendix 4.G to be

$$\hat{\sigma}_j^2 = \frac{(n-1)V_{ij}}{n}; \qquad j \neq i \qquad (4.5.1)$$

where V_{ij} is defined by Eq. 4.2.3.

Statistical inference problems related to CMLE's require no solutions not already presented in Section 1.7, since V_{ij} is a sample variance with $(n-1)$ degrees of freedom.

EXAMPLE 4.L

Eight samples of uranium-bearing materials were each measured once for percent uranium by 3 laboratories. The data are given in Table 4.5.

Applying Eq. 4.2.1 gives

$$V_{12} = 4.021455 \qquad V_{13} = 5.969130 \qquad V_{23} = 1.926619$$

From Eqs. 4.2.4 and 4.3.2, the MLE's are

$$\hat{\sigma}_1^2 = 3.5280 \qquad \hat{\sigma}_2^2 = -0.0092 \qquad \sigma_3^2 = 1.6950$$

Table 4.5   Percent Uranium for Three Laboratories

Sample	Lab 1	Lab 2	Lab 3
1	63.92	67.04	68.10
2	63.63	63.43	62.74
3	87.07	86.64	89.03
4	82.45	85.80	86.90
5	84.25	85.94	84.27
6	71.39	72.03	73.35
7	83.59	80.88	80.92
8	59.07	60.69	59.60

Since $\hat{\sigma}_2^2$ is negative, the estimate of σ_2^2 is zero. From Eq. 4.5.1, the CMLE's of σ_1^2 and σ_3^2 are then

$$\hat{\sigma}_1^2 = \frac{(7)(4.021455)}{8} = 3.5188$$

$$\hat{\sigma}_3^2 = 1.6858$$

The CMLE's are in close agreement with the MLE's in this example, since $\hat{\sigma}_2^2$ is almost zero. ∎

EXAMPLE 4.M

Returning to Example 4.A, suppose that the Method 3 data given in Table 4.1 were replaced by data for Method 4, reproduced in Table 4.6 in the same pellet order used for Methods 1 and 6 in Table 3.1. Method 4 is also an immersion method.

Applying Eq. 4.2.1 gives

$$V_{16} = 0.044502 \qquad V_{14} = 0.073115 \qquad V_{64} = 0.122010$$

From Eqs. 4.2.4 and 4.3.2, the MLE's are

$$\hat{\sigma}_1^2 = -0.002145 \qquad \hat{\sigma}_6^2 = 0.045612 \qquad \hat{\sigma}_4^2 = 0.073560$$

Since $\hat{\sigma}_1^2$ is negative, the estimate of σ_1^2 is zero. From Eq. 4.5.1, the estimates of σ_6^2 and σ_4^2 are then

$$\hat{\sigma}_6^2 = \frac{(42)(0.044502)}{43} = 0.043467$$

$$\hat{\sigma}_4^2 = 0.071415$$

Table 4.6   Pellet Densities for Method 4

4.68	4.73	4.46	4.86
5.15	4.60	4.07	4.53
4.78	4.62	4.55	4.96
5.55	4.34	4.84	4.60
5.08	5.09	4.70	4.21
4.80	5.26	4.85	4.79
4.41	5.05	4.35	4.62
5.32	4.60	5.06	5.05
4.12	5.23	4.02	4.74
4.41	5.24	4.86	4.42
4.37	4.42	4.40	

Recall from Example 4.A that $\hat{\sigma}_1^2 = 0.004086$ and $\hat{\sigma}_6^2 = 0.039382$. From Example 3.A, these estimates were 0.009783 and 0.033684, respectively.

Suppose 95% confidence limits on σ_4^2 and σ_6^2 were to be found based on the CMLE's. Then, since $28V_{14}/\sigma_4^2$ is distributed as chi-square with 28 degrees of freedom by Eq. 1.6.3, from Table 1.2,

$$\Pr\left[13.86 < \frac{(28)(0.073115)}{\sigma_4^2} < 41.90\right]$$

and the 95% confidence interval on σ_4^2 is

$$0.0489 < \sigma_4^2 < 0.1477$$

Similarly, for σ_6^2, the 95% confidence interval is

$$0.0297 < \sigma_6^2 < 0.0899 \qquad\qquad \blacksquare$$

EXAMPLE 4.N

The data for this example deal with measuring automobile fuel consumption; they were collected by the Products Research Division, Exxon Research and Engineering Company. For the purposes of this example, the data in Table 4.7 have been masked, but in such a way that the measurement parameter characteristics are unaffected. Each value is an average of five readings for a given speed and gear ratio. The measurement methods identified as Methods 1, 2, and 3 are based on carbon balance, flowmeter, and weight measurements of fuel consumption, respectively.

Applying Eq. 4.2.1 gives

$$V_{12} = 0.285031 \qquad V_{13} = 0.390947 \qquad V_{23} = 0.039604$$

From Eqs. 4.2.4 and 4.3.2, the MLE's are

$$\hat{\sigma}_1^2 = 0.2983 \qquad \hat{\sigma}_2^2 = -0.0311 \qquad \hat{\sigma}_3^2 = 0.0682$$

Since $\hat{\sigma}_2^2$ is negative, the estimate of σ_2^2 is zero. From Eq. 4.5.1, the estimates of σ_1^2 and σ_3^2 are then

$$\hat{\sigma}_1^2 = \frac{15(0.285031)}{16} = 0.2672; \qquad \hat{\sigma}_1 = 0.517 \text{ mpg}$$

$$\hat{\sigma}_3^2 = \frac{15(0.039604)}{16} = 0.0371; \qquad \hat{\sigma}_3 = 0.193 \text{ mpg} \qquad \blacksquare$$

Table 4.7   Automobile Fuel Consumption (mpg)

Method 1	Method 2	Method 3
20.960	20.630	20.634
25.570	25.684	25.628
15.088	14.406	14.276
15.694	14.978	14.498
27.876	27.372	27.160
25.890	25.480	25.268
26.036	25.904	25.752
28.374	28.920	28.474
24.328	24.414	24.474
31.570	32.120	32.290
22.776	22.308	22.026
29.174	29.914	29.934
21.018	20.348	20.176
26.898	27.250	27.464
19.690	18.570	18.614
24.790	24.636	24.670

4.6   CONSTRAINED EXPECTED LIKELIHOOD ESTIMATION (CELE)

By the reasoning discussed in Section 3.6 for $N = 2$ measurement methods, it may be desirable on occasion to find the CELE's of the parameters. Such estimates are attractive when one is reluctant to assign the value zero to a measurement error variance.

It requires more effort to find the CELE's for $N = 3$ than it did for $N = 2$. As derived in Appendix 4.H, the CELE of σ_1^2 is given by

$$\dot{\sigma}_1^2 = \frac{S \int_0^1 v w_v \, dv}{\int_0^1 w_v \, dv} \tag{4.6.1}$$

where

$$S = \frac{0.5(n - 1)(V_{12} + V_{13} + V_{23})}{n} \tag{4.6.2}$$

and

$$w_v = \exp(L_v - C) \tag{4.6.3}$$

The log likelihood L_v is given by Eq. 4.A.12 evaluated at a given set of values, σ_{1v}^2, σ_{2v}^2, and σ_{3v}^2, where, for given v between 0 and 1,

$$\sigma_{1v}^2 = vS \tag{4.6.4}$$

$$\sigma_{2v}^2 = (1 - v)S - \sigma_{3v}^2 \tag{4.6.5}$$

and where σ_{3v}^2 is the real root solution of the cubic equation

$$\sigma_{3v}^6 + a_2\sigma_{3v}^4 + a_1\sigma_{3v}^2 + a_0 = 0 \tag{4.6.6}$$

where

$$a_2 = \frac{(n - 1)(V_{12} - V_{13}) - 3n(1 - v)S}{2n} \tag{4.6.7}$$

$$a_1 = \frac{S\{n(1 - v)(1 - 3v)S + 2(n - 1)[vV_{23} + (1 - v)V_{13}]\}}{2n} \tag{4.6.8}$$

$$a_0 = \frac{(1 - v)S^2\{nv(1 - v)S + (n - 1)[v(V_{12} - V_{23}) - V_{13}]\}}{2n} \tag{4.6.9}$$

Equation 4.6.6 may be solved by using a simple computer program, a hand-calculator program, or by trial and error.

In practice, since the integrals in Eq. 4.6.1 cannot be evaluated directly, nor can explicit expressions for w_v be written as a function of v as was possible for $N = 2$, $\dot{\sigma}_1^2$ can be found by computing w_v for discrete values of v between 0 and 1 and finding the weighted average. This process is illustrated by examples. The computational process is easily programmable. The subscript 1 should be associated with the parameter whose MLE is negative, or, if all estimates are positive, is the closest to zero.

EXAMPLE 4.O

In Example 4.M, the input statistics were found to be

$$n = 43$$

$$V_{16} = 0.044502 \qquad V_{14} = 0.073115 \qquad V_{64} = 0.122010$$

The quantity S is calculated from Eq. 4.6.2:

$$S = 0.117027$$

Table 4.8　Computational Results for Example 4.O

v	σ^2_{1v}	a_2	a_1	a_0	σ^2_{3v}	σ^2_{2v}	$L_v - C$	r_v
0	0	−0.189515	0.015205	−0.000489	0.072984	0.044043	81.156	1
0.025	0.002926	−0.185126	0.014673	−0.000470	0.071997	0.042104	80.944	0.850
0.050	0.005851	−0.180738	0.014166	−0.000453	0.071111	0.040065	80.733	0.655
0.075	0.008777	−0.176349	0.013685	−0.000437	0.070043	0.038207	80.390	0.465
0.100	0.011703	−0.171961	0.013230	−0.000422	0.068822	0.036502	79.980	0.309
0.125	0.014628	−0.167572	0.012801	−0.000408	0.067470	0.034929	79.512	0.193
0.150	0.017554	−0.163184	0.012397	−0.000395	0.066033	0.033440	78.994	0.115
0.175	0.020480	−0.158795	0.012019	−0.000383	0.064516	0.032031	78.429	0.065
0.200	0.023405	−0.154407	0.011666	−0.000372	0.062960	0.030662	77.820	0.036
0.225	0.026331	−0.150018	0.011339	−0.000361	0.061132	0.029564	77.168	0.019
0.250	0.029257	−0.145630	0.011038	−0.000351	0.059306	0.028464	76.473	0.009

Table 4.8 gives computational results for given values of v. The table does not extend beyond $v = 0.25$ because w_v becomes so small in a relative sense for $v > 0.25$. The last column gives $r_v = w_v/w_0$, which for convenience may be used in place of w_v as the weights.

Then,

$$\dot{\sigma}^2_1 = \frac{(0)(1) + (0.002926)(0.850) + \cdots + (0.029257)(0.009)}{1 + 0.850 + \cdots + 0.009} = 0.005866$$

$$\dot{\sigma}^2_3 = 0.070874 \left(= \dot{\sigma}^2_4 \right)$$

$$\dot{\sigma}^2_2 = 0.040287 \left(= \dot{\sigma}^2_6 \right)$$

Recall from Example 4.M that the CMLE's were 0, 0.071415, and 0.043467, respectively.　■

EXAMPLE 4.P

In Example 4.N dealing with the measurement of automobile fuel consumption, the MLE of σ^2_2 was negative. The relevant statistics are

$$n = 16$$

$$V_{12} = 0.285031 \qquad V_{13} = 0.390947 \qquad V_{23} = 0.039604$$

From Eq. 4.6.2, $S = 0.3354$.

Table 4.9 Computational Results for Example 4.P

v	σ_{1v}^2	σ_{2v}^2	σ_{3v}^2	$L_v - C$	$r_v = w_v/w_0$
0	0.2978	0	0.0376	20.858	1
0.025	0.2958	0.0084	0.0312	20.437	0.656
0.050	0.2931	0.0168	0.0255	19.974	0.413
0.075	0.2899	0.0252	0.0204	19.480	0.252
0.100	0.2860	0.0335	0.0159	18.965	0.151
0.125	0.2816	0.0419	0.0119	18.436	0.089
0.150	0.2767	0.0503	0.0084	17.895	0.052
0.175	0.2713	0.0587	0.0054	17.348	0.030
0.200	0.2655	0.0671	0.0029	16.794	0.017
0.225	0.2591	0.0755	0.0008	16.236	0.010

Table 4.9 gives computational results for given values of v. Then,

$$\dot{\sigma}_1^2 = \frac{(0.2978)(1) + \cdots + (0.2591)(0.010)}{1 + 0.656 + \cdots + 0.010} = 0.2936; \qquad \dot{\sigma}_1 = 0.54 \text{ mpg}$$

$$\dot{\sigma}_2^2 = 0.0127; \qquad \dot{\sigma}_2 = 0.11 \text{ mpg}$$

$$\dot{\sigma}_3^2 = 0.0291; \qquad \dot{\sigma}_3 = 0.17 \text{ mpg} \qquad\qquad \blacksquare$$

4.7 ESTIMATION OF RELATIVE BIASES

In Appendix 4.A, Eq. 4.A.11, the relative bias estimate is given as

$$\widehat{\alpha_i - \alpha_j} = \frac{\sum\limits_{k=1}^{n} y_{ijk}}{n} = \bar{y}_{ij} \tag{4.7.1}$$

which is simply the mean of the difference for methods i and j.

If the α's are randomly selected from a population of α's for which the mean is zero and the variance is σ_α^2, then σ_α^2 is estimated by first calculating the variance of the method averages:

$$s_{\bar{x}}^2 = \frac{\sum\limits_{i=1}^{3} \bar{x}_i^2 - \left(\sum\limits_{i=1}^{3} \bar{x}_i\right)^2 \Big/ 3}{2} \tag{4.7.2}$$

where

$$\bar{x}_i = \frac{\sum\limits_{k=1}^{n} x_{ik}}{n} ; \qquad i = 1, 2, 3 \tag{4.7.3}$$

Then, noting that

$$E\left(s_{\bar{x}}^2\right) = \sigma_\alpha^2 + \frac{\sigma_1^2 + \sigma_2^2 + \sigma_3^2}{3n} \tag{4.7.4}$$

σ_α^2 is estimated by

$$\hat{\sigma}_\alpha^2 = s_{\bar{x}}^2 - \frac{\hat{\sigma}_1^2 + \hat{\sigma}_2^2 + \hat{\sigma}_3^2}{3n} \tag{4.7.5}$$

For purposes of calculation, it is noted that

$$s_{\bar{x}}^2 = \frac{\bar{y}_{12}^2 + \bar{y}_{13}^2 + \bar{y}_{23}^2}{6} \tag{4.7.6}$$

which may be simpler to compute in practice since the y_{ijk}'s will have already been calculated for other purposes.

EXAMPLE 4.Q

In Example 4.C, assume that the three observers are randomly selected from a population of observers and estimate σ_α^2, the variance among observers. The mean differences are found to be

$$\bar{y}_{12} = 0.02379 \qquad \bar{y}_{13} = 0.00966 \qquad \bar{y}_{23} = -0.01414$$

From Eq. 4.7.6,

$$s_{\bar{x}}^2 = 0.000367$$

From Example 4.C,

$$\hat{\sigma}_1^2 = 0.000619 \qquad \hat{\sigma}_2^2 = 0.000061 \qquad \hat{\sigma}_3^2 = 0.000240$$

Therefore, from Eq. 4.7.5, with $n = 29$,

$$\hat{\sigma}_\alpha^2 = 0.000356 \text{ sec}^2$$

$$\hat{\sigma}_\alpha = 0.0189 \text{ sec} \qquad\qquad \blacksquare$$

DERIVATION OF MAXIMUM LIKELIHOOD ESTIMATES

First, the joint distribution of y_{12k} and y_{13k} is considered. From Eq. 4.1.2,

$$E(y_{1jk}) = (\alpha_1 - \alpha_j); \qquad j = 2,3 \tag{4.A.1}$$

$$E(y_{1jk} - \alpha_1 + \alpha_j)^2 = \sigma_1^2 + \sigma_j^2; \qquad j = 2,3 \tag{4.A.2}$$

$$E(y_{12k} - \alpha_1 + \alpha_2)(y_{13k} - \alpha_1 + \alpha_3) = \sigma_1^2 \tag{4.A.3}$$

Thus, the variance-covariance matrix is

$$(\sigma_{ij}) = \begin{pmatrix} \sigma_1^2 + \sigma_2^2 & \sigma_1^2 \\ \sigma_1^2 & \sigma_1^2 + \sigma_3^2 \end{pmatrix} \tag{4.A.4}$$

The determinant of (σ_{ij}), denoted by Q, is

$$Q = \sigma_1^2\sigma_2^2 + \sigma_1^2\sigma_3^2 + \sigma_2^2\sigma_3^2 \tag{4.A.5}$$

and, hence, the inverse matrix is

$$(\sigma^{ij}) = Q^{-1} \begin{pmatrix} \sigma_1^2 + \sigma_3^2 & -\sigma_1^2 \\ -\sigma_1^2 & \sigma_1^2 + \sigma_2^2 \end{pmatrix} \tag{4.A.6}$$

The likelihood for y_{12k} and y_{13k} may now be written, its density function being the bivariate normal density:

$$f(y_{12k}, y_{13k}) = \frac{1}{2\pi\sqrt{Q}} \exp\left\{ \frac{-0.5}{Q} \left[(\sigma_1^2 + \sigma_3^2)(y_{12k} - \alpha_1 + \alpha_2)^2 \right.\right.$$

$$- 2\sigma_1^2(y_{12k} - \alpha_1 + \alpha_2)(y_{13k} - \alpha_1 + \alpha_3)$$

$$\left.\left. + (\sigma_1^2 + \sigma_2^2)(y_{13k} - \alpha_1 + \alpha_3)^2 \right] \right\} \tag{4.A.7}$$

Now, note that since

$$y_{23k} - \alpha_2 + \alpha_3 = (y_{13k} - \alpha_1 + \alpha_3) - (y_{12k} - \alpha_1 + \alpha_2) \qquad (4.A.8)$$

then

$$2(y_{12k} - \alpha_1 + \alpha_2)(y_{13k} - \alpha_1 + \alpha_3)$$

$$= (y_{12k} - \alpha_1 + \alpha_2)^2 + (y_{13k} - \alpha_1 + \alpha_3)^2$$

$$- (y_{23k} - \alpha_2 + \alpha_3)^2 \qquad (4.A.9)$$

and hence, the expression in the square brackets in Eq. 4.A.7 may be rewritten:

$$[\ldots] = \sigma_3^2(y_{12k} - \alpha_1 + \alpha_2)^2 + \sigma_2^2(y_{13k} - \alpha_1 + \alpha_3)^2 + \sigma_1^2(y_{23k} - \alpha_2 + \alpha_3)^2$$

$$(4.A.10)$$

For the n observations, the likelihood is found from Eq. 4.A.7 by raising the coefficient $(2\pi\sqrt{Q})^{-1}$ to the nth power and summing the exponent for $k = 1$ to n. With $(\alpha_i - \alpha_j)$ estimated by

$$\widehat{\alpha_i - \alpha_j} = \frac{\sum\limits_{k=1}^{n} y_{ijk}}{n} \qquad (4.A.11)$$

and with V_{ij} defined by Eq. 4.2.3, the natural logarithm of the likelihood function, denoted by L, is now

$$L = -n \ln 2\pi - 0.5n \ln Q$$

$$- 0.5(n - 1)Q^{-1}(\sigma_3^2 V_{12} + \sigma_2^2 V_{13} + \sigma_1^2 V_{23}) \qquad (4.A.12)$$

The partial derivative of L with respect to σ_1^2 is found and equated to zero, keeping in mind that Q is also a function of σ_1^2 and that

$$\frac{\partial Q}{\partial \sigma_1^2} = \sigma_2^2 + \sigma_3^2 \qquad (4.A.13)$$

$$\frac{\partial L}{\partial \sigma_1^2} = \frac{-0.5n(\sigma_2^2 + \sigma_3^2)}{Q} - \frac{0.5(n - 1)(\text{NUM})}{Q^2} = 0 \qquad (4.A.14)$$

where

$$\text{NUM} = -\sigma_3^2\left(\sigma_2^2 + \sigma_3^2\right)V_{12} - \sigma_2^2\left(\sigma_2^2 + \sigma_3^2\right)V_{13} + \sigma_2^2\sigma_3^2 V_{23} \quad (4.A.15)$$

Upon multiplying both sides of Eq. 4.A.14 by $-2Q^2$ and solving for σ_1^2, the solution is

$$\sigma_1^2 = \frac{-(n-1)(\text{NUM})}{n\left(\sigma_2^2 + \sigma_3^2\right)^2} - \frac{\sigma_2^2\sigma_3^2}{\sigma_2^2 + \sigma_3^2} \quad (4.A.16)$$

which easily reduces to Eq. 4.3.1.

PROOF THAT MAXIMUM LIKELIHOOD ESTIMATES ARE $(n - 1)/n$ TIMES GRUBBS' ESTIMATORS

To prove that $\hat{\sigma}_i^2$ given by Eq. 4.3.1 is equal to $(n - 1)\,\tilde{\sigma}_i^2/n$, with $\tilde{\sigma}_i^2$ given by Eq. 4.2.4, replace $\hat{\sigma}_i^2$ in Eq. 4.3.1 by $(n - 1)\,\tilde{\sigma}_i^2/n$ and show that Eq. 4.3.1 is satisfied.

The left-hand side (LHS) of Eq. 4.3.1 becomes

$$\text{LHS} = \frac{0.5(n - 1)(V_{12} + V_{13} - V_{23})}{n} \tag{4.B.1}$$

In simplifying the right-hand side (RHS), note from Eq. 4.2.3 that

$$\sigma_2^2 + \sigma_3^2 = \frac{n - 1}{n} V_{23} \tag{4.B.2}$$

Further,

$$\sigma_2^2 \sigma_3^2 = \frac{(n - 1)^2}{4n^2} - \left(V_{12}^2 - V_{13}^2 + V_{23}^2 + 2V_{12}V_{13}\right) \tag{4.B.3}$$

and

$$\sigma_3^2 V_{12} + \sigma_2^2 V_{13} = \frac{n - 1}{2n}\left(-V_{12}^2 - V_{13}^2 + 2V_{12}V_{13} + V_{12}V_{23} + V_{13}V_{23}\right) \tag{4.B.4}$$

The RHS of Eq. 4.3.1 is now rewritten to give

RHS

$$= \frac{(n - 1)\left(\sigma_2^2 + \sigma_3^2\right)\left(\sigma_3^2 V_{12} + \sigma_2^2 V_{13}\right) - (n - 1)\sigma_2^2 \sigma_3^2 V_{23} - n\sigma_2^2 \sigma_3^2\left(\sigma_2^2 + \sigma_3^2\right)}{n\left(\sigma_2^2 + \sigma_3^2\right)^2} \tag{4.B.5}$$

Equations 4.B.2, 4.B.3 and 4.B.4 are now applied to the numerator of Eq. 4.B.5 to give

$$
\text{NUM} = \frac{(n-1)^3}{2n^2} \left(-V_{12}^2 V_{23} - V_{13}^2 V_{23} + 2V_{12}V_{13}V_{23} + V_{12}V_{23}^2 + V_{13}V_{23}^2 \right)
$$

$$
- \frac{(n-1)^3}{4n^2} \left(-V_{12}^2 V_{23} - V_{13}^2 V_{23} + V_{23}^3 + 2V_{12}V_{13}V_{23} \right)
$$

$$
- \frac{(n-1)^3}{4n^2} \left(-V_{12}^2 V_{23} - V_{13}^2 V_{23} + V_{23}^3 + 2V_{12}V_{13} \right)
$$

$$
= \frac{(n-1)^3 V_{23}^2}{2n^2} \left(V_{12} + V_{13} - V_{23} \right) \tag{4.B.6}
$$

The denominator of Eq. 4.B.5 is, from Eq. 4.B.2,

$$
\text{DEN} = \frac{(n-1)^2}{n} V_{23}^2 \tag{4.B.7}
$$

Therefore, from Eqs. 4.B.6 and 4.B.7,

$$
\text{RHS} = \frac{0.5(n-1)(V_{12} + V_{13} - V_{23})}{n} \tag{4.B.8}
$$

which is identical to the LHS of Eq. 4.B.1, and the proof is completed.

DERIVATION OF CONFIDENCE INTERVALS BASED ON LARGE-SAMPLE THEORY

The element in position (i, i) of the H matrix, Eq. 4.4.3, is the negative of the second partial derivative of L with respect to σ_i^2. The element in position (i, j) is the negative of the partial derivative of $(\partial L/\partial \sigma_i^2)$ with respect to σ_j^2.

For convenience in notation, replace σ_i^2 by v_i. Then, L in Eq. 4.A.12 may be written in the form

$$L = C - 0.5n \ln Q - \frac{0.5(n-1)A}{Q} \tag{4.C.1}$$

where

$$Q = v_1 v_2 + v_1 v_3 + v_2 v_3 \tag{4.C.2}$$

and

$$A = v_3 V_{12} + v_2 V_{13} + v_1 V_{23} \tag{4.C.3}$$

Then,

$$\frac{\partial L}{\partial v_1} = \frac{-0.5n(v_2 + v_3)}{Q} \frac{-0.5(n-1)[QV_{23} - A(v_2 + v_3)]}{Q^2} \tag{4.C.4}$$

and $(\partial^2 L/\partial v_1^2)$ reduces to

$$\frac{\partial^2 L}{\partial v_1^2} = \frac{(v_2 + v_3)^2}{Q^2} \left[0.5n + \frac{(n-1)V_3}{v_2 + v_3} - \frac{(n-1)A}{Q} \right] \tag{4.C.5}$$

It is easily shown from the argument in Appendix 4.B that

$$\frac{(n-1)V_{23}}{v_2 + v_3} = n \tag{4.C.6}$$

and that

$$\frac{(n-1)A}{Q} = 2n \tag{4.C.7}$$

so that

$$\frac{\partial^2 L}{\partial v_1^2} = \frac{-0.5n(v_2 + v_3)^2}{Q^2} \tag{4.C.8}$$

By interchanging appropriate subscripts, the diagonal elements in the H matrix of Eq. 4.4.3 have been derived.

Next, by differentiating Eq. 4.C.4 with respect to v_2, the result is

$$\frac{\partial^2 L}{\partial v_1 \partial v_2} = \frac{0.5nv_3^2}{Q^2} + \frac{0.5(n-1)}{Q^2}\left[(v_1 + v_3)V_{23} + (v_2 + v_3)V_{13} + A\right]$$

$$- \frac{(n-1)A(v_1 + v_3)(v_2 + v_3)}{Q^3} \tag{4.C.9}$$

By expressing A, Q, and v_i as functions of V_{12}, V_{13}, and V_{23}, it follows that the last two expressions in Eq. 4.C.9 sum to minus twice the first term. Thus

$$\frac{\partial^2 L}{\partial v_1 \partial v_2} = \frac{-0.5nv_3^2}{Q^2} \tag{4.C.10}$$

By interchanging appropriate subscripts, the off-diagonal elements in the H matrix of Eq. 4.4.3 have been derived.

DERIVATION OF $L(\hat{\Omega})$, EQUATION 4.4.5

In Eq. 4.A.5, replace the σ_i^2's by their respective MLE's given by Eqs. 4.2.4 and 4.3.2:

$$Q = \frac{0.25(n-1)^2\left(-V_{12}^2 - V_{13}^2 - V_{23}^2 + 2V_{12}V_{13} + 2V_{12}V_{23} + 2V_{13}V_{23}\right)}{n^2}$$

$$(4.D.1)$$

In Eq. 4.A.12, the expression $(\sigma_3^2 V_{12} + \sigma_2^2 V_{13} + \sigma_1^2 V_{23})$ reduces to

$$\frac{0.5(n-1)\left(-V_{12}^2 - V_{13}^2 - V_{23}^2 + 2V_{12}V_{13} + 2V_{12}V_{23} + 2V_{13}V_{23}\right)}{n}$$

and upon dividing this by Q of Eq. 4.D.1, L becomes

$$L(\hat{\Omega}) = -n\ln 2\pi - 0.5n\ln Q - n \qquad (4.D.2)$$

which is Eq. 4.4.5, where $C = -n\ln 2\pi$.

DERIVATION OF MAXIMUM LIKELIHOOD ESTIMATORS OF σ_0^2 AND OF $L(\hat{\omega})$ WHEN PRECISIONS ARE EQUAL

Under H_{01}: $\sigma_1^2 = \sigma_2^2 = \sigma_3^2 \; (= \sigma_0^2)$, L of equation Eq. 4.A.12 reduces to

$$L = -n \ln 2\pi - 0.5n \ln 3\sigma_0^4 - \frac{0.5(n-1)\sigma_0^2(V_{12} + V_{13} + V_{23})}{3\sigma_0^4}$$

$$= -n \ln 2\pi - 0.5n \ln 3 - 0.5n \ln(\sigma_0^2)^2 - \frac{0.5(n-1)(V_{12} + V_{13} + V_{23})}{3\sigma_0^2}$$

$$\text{(4.E.1)}$$

Then,

$$\frac{\partial L}{\partial \sigma_0^2} = \frac{-n}{\sigma_0^2} + \frac{0.5(n-1)(V_{12} + V_{13} + V_{23})}{3\sigma_0^4} = 0$$

from which

$$\hat{\sigma}_0^2 = \frac{(n-1)(V_{12} + V_{13} + V_{23})}{6n} \qquad \text{(4.E.2)}$$

which is Eq. 4.4.6.

Inserting this value in Eq. 4.E.1 yields

$$L(\hat{\omega}) = -n \ln 2\pi - 0.5n\left(\ln 3 + 2 \ln \hat{\sigma}_0^2\right) - n \qquad \text{(4.E.3)}$$

which is Eq. 4.4.6 with $C = -n \ln 2\pi$.

PROOF THAT NO MORE THAN ONE
PRECISION ESTIMATE CAN BE NEGATIVE

Without loss of generality, assume that $\tilde{\sigma}_1^2$ (or $\hat{\sigma}_1^2$) is negative. This means that

$$V_{12} + V_{13} - V_{23} < 0 \qquad (4.F.1)$$

or, equivalently,

$$V_{23} = V_{12} + V_{13} + \Delta \qquad (4.F.2)$$

where Δ is positive. Equation 4.F.2 then implies that both $(V_{12} - V_{13} + V_{23})$ and $(-V_{12} + V_{13} + V_{23})$ are positive, and hence $\hat{\sigma}_2^2$ and $\hat{\sigma}_3^2$ are positive. Specifically,

$$V_{12} - V_{13} + V_{23} = V_{12} - V_{13} + V_{12} + V_{13} + \Delta$$

$$= 2V_{12} + \Delta \qquad (4.F.3)$$

which is a positive quantity since V_{12} is nonnegative and Δ is positive. Similarly, $(-V_{12} + V_{13} + V_{23})$ is also positive, thus completing the proof.

DERIVATION OF CONSTRAINED MAXIMUM LIKELIHOOD ESTIMATES (CMLE'S)

Without loss of generality, assume that $\hat{\sigma}_1^2 < 0$. Then, in the likelihood function L, given by Eq. 4.A.12, σ_1^2 is set equal to 0, which is the value of σ_1^2 that maximizes L in the space of nonnegative values of σ_1^2, and L becomes

$$L = -n\ln 2\pi - 0.5n\left(\ln\sigma_2^2 + \ln\sigma_3^2\right)$$

$$-0.5(n-1)\left(\frac{V_{12}}{\sigma_2^2} + \frac{V_{13}}{\sigma_3^2}\right) \tag{4.G.1}$$

Then,

$$\frac{\partial L}{\partial \sigma_2^2} = -\frac{0.5n}{\sigma_2^2} + \frac{0.5(n-1)V_{12}}{\sigma_2^4} = 0 \tag{4.G.2}$$

from which

$$\hat{\sigma}_2^2 = \frac{(n-1)V_{12}}{n} \tag{4.G.3}$$

Clearly, the derivation of the CMLE for σ_3^2 is identical, thus establishing the validity of Eq. 4.5.1.

DERIVATION OF CONSTRAINED EXPECTED LIKELIHOOD ESTIMATES (CELE'S)

Equations 4.2.4 and 4.3.2 indicate that the MLE of $(\sigma_1^2 + \sigma_2^2 + \sigma_3^2)$ is S:

$$S = \frac{0.5(n-1)(V_{12} + V_{13} + V_{23})}{n} \qquad (4.H.1)$$

If the CELE's are denoted by $\dot{\sigma}_1^2$, $\dot{\sigma}_2^2$, and $\dot{\sigma}_3^2$, respectively, they will by definition sum to S. The CELE of σ_1^2 is found by calculating the log likelihood function L, given by Eq. 4.A.12, corresponding to

$$\sigma_{1v}^2 = vS \qquad (4.H.2)$$

and

$$\sigma_{2v}^2 + \sigma_{3v}^2 = (1-v)S \qquad (4.H.3)$$

where, for given v, σ_{3v}^2 is chosen to maximize L. Denote the value of L that is found under the above conditions by L_v. Then, σ_1^2 is the weighted average

$$\dot{\sigma}_1^2 = \frac{S \int_0^1 v w_v \, dv}{\int_0^1 w_v \, dv} \qquad (4.H.4)$$

where

$$w_v = \exp(L_v - C) \qquad (4.H.5)$$

In order to find $\dot{\sigma}_1^2$, it is first necessary to find σ_{3v}^2. From Eqs. 4.A.12, 4.H.2, and 4.H.3, the log likelihood function for given v is of the form

$$L = C - 0.5n \ln Q - \frac{0.5(n-1)A}{Q} \qquad (4.H.6)$$

where, from Eqs. 4.A.5, 4.H.2, and 4.H.3,

$$Q = v(1 - v)S^2 + (1 - v)S\sigma_{3v}^2 - \sigma_{3v}^4 \qquad (4.\text{H}.7)$$

and

$$A = \sigma_{3v}^2(V_{12} - V_{13}) + vS(V_{23} - V_{13}) + SV_{13} \qquad (4.\text{H}.8)$$

σ_{3v}^2 is chosen to maximize L for given v. By taking the partial derivative of L with respect to σ_{3v}^2 and equating it to zero, the resulting equation is

$$Q[nB + (n - 1)D] - (n - 1)AB = 0 \qquad (4.\text{H}.9)$$

where

$$B = \frac{\partial Q}{\partial \sigma_{3v}^2} = (1 - v)S - 2\sigma_{3v}^2 \qquad (4.\text{H}.10)$$

and

$$D = \frac{\partial A}{\partial \sigma_{3v}^2} = V_{12} - V_{13} \qquad (4.\text{H}.11)$$

Equation 4.H.9 is a cubic equation in σ_{v3}^2 of the form:

$$\sigma_{3v}^6 + a_2\sigma_{3v}^4 + a_1\sigma_{3v}^2 + a_0 = 0 \qquad (4.\text{H}.12)$$

where

$$a_2 = \frac{(n - 1)(V_{12} - V_{13}) - 3n(1 - v)S}{2n} \qquad (4.\text{H}.13)$$

$$a_1 = \frac{S\{n(1 - v)(1 - 3v)S + 2(n - 1)[vV_{23} + (1 - v)V_{13}]\}}{2n}$$

$$(4.\text{H}.14)$$

$$a_0 = \frac{(1 - v)S^2\{nv(1 - v)S + (n - 1)[v(V_{12} - V_{23}) - V_{13}]\}}{2n}$$

$$(4.\text{H}.15)$$

For given n, V_{12}, V_{13}, V_{23}, and v, Eq. 4.H.12 can be solved for σ_{3v}^2. Then σ_{1v}^2 and σ_{2v}^2 are computed from Eqs. 4.H.2 and 4.H.3 and L_v is computed from Eq. 4.A.12 using these values of the parameters.

REFERENCES

1. Grubbs, F. E. "On Estimating Precision of Measuring Instruments and Product Variability." *J. Amer. Stat. Assn.* **43**: 243–264; 1948.

2. Mandel, J. "Models, Transformations of Scale, and Weighting." *J. Qual. Technol.* **8**(2): 86–97; 1976.

3. Jaech, J. L. "Estimating Within-Laboratory Variability from Interlaboratory Test Data." *J. Qual. Technol.* **11**(4): 185–191; 1979; Table 2.

4. Hahn, G. J. and Shapiro, S. S. *Statistical Models in Engineering.* New York: John Wiley and Sons; 1967; p. 128.

5. Mood, A. M. *Introduction to the Theory of Statistics.* New York: McGraw-Hill; 1950.

6. Jaech, J. L. "Large Sample Tests for Grubbs' Estimators of Instrument Precision with More Than Two Instruments." *Technometrics.* **18**(2): 127–133; 1976.

7. Russell, T. S. and Bradley, R. A. "One-Way Variance in a Two-Way Classification." *Biometrika.* **45**: 111–129; 1958.

Chapter 5

$N \geq 4$ MEASUREMENT METHODS: INFERENCE BASED ON PAIRED DIFFERENCES

5.1 INTRODUCTION

With reference to the model given as Eq. 1.8.1, in this chapter we study the case of $N \geq 4$ measurement methods with $\beta_1 = \beta_2 = \cdots = \beta_N$; $\alpha_1 \neq \alpha_2 \neq \cdots \neq \alpha_N$; and $\sigma_1^2 \neq \sigma_2^2 \neq \cdots \neq \sigma_N^2$. The problem formulation is quite similar to that for $N = 3$ measurement methods, and the outline of Chapter 4 for $N = 3$ is generally followed also for $N \geq 4$. However, because of complexity and because negative estimates of σ_i^2 are uncommon for $N \geq 4$, constrained expected likelihood estimates (CELE's) are not included in Chapter 5. Chapter 5 becomes easier to read if the parallel results for $N = 3$ are kept in mind.

Equations 4.1.1 and 4.1.2 also apply in Chapter 5, with the index i in Eq. 4.1.1 now going from 1 to N rather than 1 to 3. As in Chapter 4, no distinction need be made between the random and fixed models.

5.2 GRUBBS' ESTIMATORS

As for $N = 3$, the columns of differences are formed and the variances V_{ij} are calculated as in Eq. 4.2.1. For general N, there are $N(N-1)/2$ V_{ij} values, with $i = 1, 2, \ldots, N-1$ and $j > i$. Also, Eq. 4.2.2 still holds, but the moments, Eq. 4.2.3, are now written

$$V_{ij} = \sigma_i^2 + \sigma_j^2; \qquad i = 1, 2, \ldots, N-1 \qquad j > i \qquad (5.2.1)$$

Now, there are $N(N-1)/2$ equations in N unknowns so that for $N \geq 4$, there are more equations than there are unknowns. For $N = 3$, recall that there were three equations in three unknowns, and Grubbs' estimators were the solutions of these equations. With more equations than there are unknowns, Grubbs' estimators are the least squares solutions of these equations, that is, $\tilde{\sigma}_i^2$ $(i = 1, 2, \ldots, N)$ is chosen to minimize the sum of squares:

$$S = \sum_{i=1}^{N-1} \sum_{j>i} \left(V_{ij} - \tilde{\sigma}_i^2 - \tilde{\sigma}_j^2 \right)^2 \qquad (5.2.2)$$

It is shown in Appendix 5.A that the Grubbs' estimate of σ_i^2 is

$$\tilde{\sigma}_i^2 = \frac{(N-1)S_i - V_T}{(N-1)(N-2)} \tag{5.2.3}$$

where

$$S_i = \sum_{\substack{j=1 \\ j \neq i}}^{N} V_{ij} \tag{5.2.4}$$

and where

$$V_T = \sum_{i=1}^{N-1} \sum_{j>i} V_{ij} = \frac{\sum_{i=1}^{N} S_i}{2} \tag{5.2.5}$$

The estimator Eq. 5.2.3 is written in a form that differs from that given by Grubbs (1), but it is identical to Grubbs' estimate, Eqs. 24 or 28 of Reference 1, where Grubbs' estimate is given for $i = 1$.

Note from Eqs. 5.2.3 to 5.2.5 that if $N = 3$ and, say, $i = 2$, then

$$\tilde{\sigma}_2^2 = \frac{2(V_{12} + V_{23}) - (V_{12} + V_{13} + V_{23})}{2} = 0.5(V_{12} - V_{13} + V_{23})$$

which is the second of Eqs. 4.2.4, illustrating that Eqs. 5.2.3 to 5.2.5 apply to all $N \geq 3$.

5.3   MAXIMUM LIKELIHOOD ESTIMATORS (MLE'S)

The MLE of σ_i^2 is derived in Appendix 5.B. An iterative estimation procedure is required, since the MLE of σ_i^2 is written as a function of the remaining σ_j^2 values. Specifically, the MLE of σ_i^2 is

$$\hat{\sigma}_i^2 = \frac{(n-1)(b_{0i}b_{1i} - b_{2i})}{nb_{0i}^2} - \frac{1}{b_{0i}} \tag{5.3.1}$$

where

$$b_{0i} = \sum_{j \neq i} \frac{1}{\sigma_j^2} \tag{5.3.2}$$

$$b_{1i} = \sum_{j \neq i} \frac{V_{ij}}{\sigma_j^2} \tag{5.3.3}$$

$$b_{2i} = \sum_{j} \sum_{k>j} \frac{V_{jk}}{\sigma_j^2 \sigma_k^2} \atop {j, k \neq i} \tag{5.3.4}$$

Note that for $i = 1$ and $N = 3$,

$$b_{01} = \frac{1}{\sigma_2^2} + \frac{1}{\sigma_3^2}$$

$$b_{11} = \frac{V_{12}}{\sigma_2^2} + \frac{V_{13}}{\sigma_3^2}$$

$$b_{21} = \frac{V_{23}}{\sigma_2^2 \sigma_3^2}$$

and Eq. 5.3.1 becomes

$$\hat{\sigma}_1^2 = \frac{(n-1)\left[(\sigma_2^2 + \sigma_3^2)(\sigma_3^2 V_{12} + \sigma_2^2 V_{13}) - \sigma_2^2 \sigma_3^2 V_{23}\right]}{n(\sigma_2^2 + \sigma_3^2)^2} - \frac{\sigma_2^2 \sigma_3^2}{\sigma_2^2 + \sigma_3^2} \tag{5.3.5}$$

which is Eq. 4.3.1, the MLE of σ_1^2 for $N = 3$. Recall, however, that for $N = 3$, an explicit solution for $\hat{\sigma}_i^2$ was found, as given by Eq. 4.3.2. Such is not the case for $N \geq 4$, where an iterative estimation procedure must be used. The procedure is to assign initial values to $\sigma_1^2, \sigma_2^2, \ldots, \sigma_{N-1}^2$ and use Eqs. 5.3.1 to 5.3.4 to determine an initial value for σ_N^2. This value and the initially assigned values for $\sigma_2^2, \ldots, \sigma_{N-1}^2$ are then used with Eqs. 5.3.1 to 5.3.4 to determine the initial calculated (as compared to assigned) value for σ_1^2. This process is continued until the calculated values at a given iteration agree, to specified criteria, with those calculated at the previous iteration. The estimation process is easily programmed, and a Fortran program computer listing is given in Chapter 9. The estimation process is illustrated in the following examples.

EXAMPLE 5.A

In Examples 3.A, 4.A, and 4.M, data are given for measurement methods 1, 3, 4, and 6 in an experiment dealing with measuring the densities of 43 cylindrical sintered uranium fuel pellets for use in a nuclear reactor, the densities having been measured by each of 6 different methods. The data [in % theoretical density (%TD) − 90%] for methods 2 and 5 are given in Table 5.1 to complete the data set. The pellet order is the same as in previous tables.

For all six methods, the V_{ij}'s from Eq. 4.2.1 are as follows, with units of 10^{-6} (%TD)^2.

$$
\begin{array}{lll}
V_{12} = 15{,}412 & V_{23} = 35{,}716 & V_{35} = 118{,}553 \\
V_{13} = 30{,}374 & V_{24} = 96{,}730 & V_{36} = 66{,}511 \\
V_{14} = 73{,}115 & V_{25} = 142{,}260 & V_{45} = 144{,}975 \\
V_{15} = 116{,}388 & V_{26} = 27{,}193 & V_{46} = 122{,}010 \\
V_{16} = 44{,}502 & V_{34} = 91{,}336 & V_{56} = 172{,}345
\end{array}
$$

The Grubbs' estimates are calculated using Eqs. 5.2.3 to 5.2.5. V_T in Eq. 5.2.5 is the sum of the 15 V_{ij} values:

$$ V_T = 1.297420 $$

Table 5.1   Pellet Densities for Methods 2 and 5 (%TD − 90%)

Method 2	Method 5	Method 2	Method 5	Method 2	Method 5
4.06	5.22	4.47	5.44	4.32	4.30
4.35	4.45	4.81	4.86	4.25	4.81
4.52	4.79	4.34	4.62	4.60	4.17
4.67	5.10	4.24	4.27	4.38	4.98
4.70	4.82	4.37	5.04	4.35	5.36
4.30	4.47	4.52	5.36	4.40	4.92
4.47	4.27	4.37	4.66	4.34	4.59
4.59	4.95	4.23	4.48	4.24	4.39
4.05	4.48	3.98	4.69	4.52	4.85
4.78	4.75	4.36	4.73	4.12	4.45
4.25	4.06	4.30	4.56	4.47	4.29
4.55	5.14	4.52	4.20	4.21	5.24
4.12	4.22	4.33	4.26	4.52	5.06
3.95	4.49	4.14	4.39		
4.27	4.29	4.59	5.50		

From repeated application of Eq. 5.2.4,

$$S_1 = (15,412 + 30,374 + \cdots + 44,502) \times 10^{-6} = 0.279791$$

$$S_2 = (15,412 + 35,716 + \cdots + 27,193) \times 10^{-6} = 0.317311$$

$$S_3 = 0.342490 \qquad S_4 = 0.528166 \qquad S_5 = 0.694521 \qquad S_6 = 0.432561$$

As a check, the S's should sum to twice V_T.
From repeated application of Eq. 5.2.3, the Grubbs' estimates are

$$\tilde{\sigma}_1^2 = \frac{(5)0.279791 - 1.297420}{20} = 0.005077$$

$$\tilde{\sigma}_2^2 = \frac{(5)0.317311 - 1.297420}{20} = 0.014457$$

$$\tilde{\sigma}_3^2 = 0.020752 \qquad \tilde{\sigma}_4^2 = 0.067171 \qquad \tilde{\sigma}_5^2 = 0.108759 \qquad \tilde{\sigma}_6^2 = 0.043269$$

Recall from Example 4.A, where $N = 3$, that the Grubbs' estimates of σ_1^2, σ_3^2, and σ_6^2 were, respectively, 0.004183, 0.026192, and 0.040320 units squared.

The MLE's are now found. The iteration process has been programmed for computer application, and the results in Table 5.2 were obtained using the computer program listed in Chapter 9.

In comparing the MLE's with the Grubbs' estimates, note the considerable differences. The ratios of the MLE's to the Grubbs' estimates are in Table 5.3.

The lack of agreement between the Grubbs' estimates and MLE's is considered later. ∎

Table 5.2 Parameter Estimates for Example 5.A in Each Iteration $(\%TD)^2$

Iteration	σ_1^2	σ_2^2	σ_3^2	σ_4^2	σ_5^2	σ_6^2
Input*	0.005000	0.014000	—	0.067000	0.109000	0.043000
1	0.004805	0.009571	0.034866	0.070000	0.112791	0.034866
2	0.005747	0.008930	0.024575	0.071728	0.113551	0.033440
3	0.006021	0.008726	0.024460	0.072439	0.113838	0.032360
⋮	⋮	⋮	⋮	⋮	⋮	⋮
8	0.006151	0.008628	0.024502	0.072837	0.113985	0.031919
9	0.006151	0.008628	0.024503	0.072838	0.113986	0.031918

*The initial values were Grubbs' estimates rounded to three decimals.

Table 5.3   Ratio of $\hat{\sigma}_i^2/\tilde{\sigma}_i^2$

i	Ratio	i	Ratio
1	1.21	4	1.08
2	0.60	5	1.05
3	1.18	6	0.74

EXAMPLE 5.B

Example 4.D is continued. This example, reported by Mandel (2), concerns the measurement of stress at 600% elongation of 7 rubber samples. Example 4.D concerned measurements made by Laboratories 6, 8, and 12, with the data listed in Table 4.4. We now include Laboratories 3 and 5. As before, the data are first transformed to natural logarithms to stabilize the variance. The data for Laboratories 3 and 5 are given in Table 5.4.

For all five methods, the V_{ij}'s from Eq. 4.2.1 are as follows on the transformed scale. All numbers are multiplied by 10^6.

$$V_{35} = 2758 \qquad V_{58} = 8256$$
$$V_{36} = 5091 \qquad V_{5,12} = 3692$$
$$V_{38} = 7083 \qquad V_{68} = 1742$$
$$V_{3,12} = 1001 \qquad V_{6,12} = 5448$$
$$V_{56} = 4512 \qquad V_{8,12} = 5153$$

The Grubbs' estimates are calculated using Eqs. 5.2.3 to 5.2.5. From Eq. 5.2.5,

$$V_T = 0.044736$$

Table 5.4   Logarithms of Stress of Rubber Samples (kg/cm^2)

Sample	Lab 3	Lab 5
1	4.209	4.033
2	4.792	4.690
3	3.419	3.334
4	4.063	3.885
5	3.332	3.107
6	3.327	3.115
7	3.817	3.665

Table 5.5 Parameter Estimates for Example 5.B in Each Iteration*

Iteration	σ_3^2	σ_5^2	σ_6^2	σ_8^2	σ_{12}^2
Input	0.001600	0.002700	—	0.003700	0.001400
1	0.000685	0.002000	0.002374	0.003233	0.000838
2	0.000507	0.002108	0.002970	0.004119	0.000577
3	0.000481	0.002142	0.003366	0.004349	0.000534
⋮	⋮	⋮	⋮	⋮	⋮
7	0.000477	0.002146	0.003442	0.004386	0.000531
8	0.000477	0.002146	0.003442	0.004386	0.000531

*See the note in Example 4.D that discusses the interpretation of these variances.

From repeated application of Eq. 5.2.4,

$$S_3 = (2758 + 5091 + 7083 + 1001) \times 10^{-6} = 0.015933$$

$$S_5 = 0.019218 \qquad S_6 = 0.016793 \qquad S_8 = 0.022234 \qquad S_{12} = 0.015294$$

From repeated application of Eq. 5.2.3, the Grubbs' estimates are

$$\tilde{\sigma}_3^2 = \frac{4(0.015933) - 0.044736}{12} = 0.001583$$

$$\tilde{\sigma}_5^2 = 0.002678 \qquad \tilde{\sigma}_6^2 = 0.001870 \qquad \tilde{\sigma}_8^2 = 0.003683 \qquad \tilde{\sigma}_{12}^2 = 0.001370$$

The MLE's are now found, with the results of each iteration displayed in Table 5.5.

The ratios of the MLE's to the Grubbs' estimates are listed in Table 5.6. Again, note the considerable differences. Comparing $\hat{\sigma}_6^2$, $\hat{\sigma}_8^2$, and $\hat{\sigma}_{12}^2$ from Example 5.B with the same estimates from Example 4.D, which involved only three laboratories, shows that inclusion of the additional two laboratories

Table 5.6 Ratio of $\hat{\sigma}_i^2 / \tilde{\sigma}_i^2$

i	Ratio
3	0.30
5	0.80
6	1.84
8	1.19
12	0.39

appreciably changes the estimates of these three parameters. This fact emphasizes the need to examine the quality of the estimates from the viewpoint of statistical inference, the topic to be considered in Section 5.4.   ■

EXAMPLE 5.C

Measurements of uranium concentration are made on each of eight samples by each of four laboratories in an interlaboratory experiment. The data for the first three laboratories have been given in Table 4.5 of Example 4.L, where it is found that the MLE of σ_2^2 is negative. The data for Laboratory 4 are given in Table 5.7. (As a note of interest, this is the data set that prompted my detailed investigation into methods of estimation for the general model, when I noted that the value of the likelihood function evaluated for Grubbs' estimates was actually smaller than the corresponding value evaluated under the hypothesis of equal precisions.)

From Eq. 4.2.1, the V_{ij}'s are

$$V_{12} = 4.021455 \qquad V_{23} = 1.926619$$

$$V_{13} = 5.969130 \qquad V_{24} = 2.654039$$

$$V_{14} = 9.258080 \qquad V_{34} = 1.613654$$

From Eqs. 5.2.4 and 5.2.5,

$$S_1 = 19.248665 \qquad S_2 = 8.602113 \qquad S_3 = 9.509403 \qquad S_4 = 13.525773$$

$$V_T = 25.442977$$

Table 5.7   Uranium Concentrations for
Eight Samples (%U)

Sample	Lab 4
1	70.36
2	64.10
3	87.43
4	86.41
5	84.27
6	73.63
7	79.70
8	59.84

Table 5.8   Parameter Estimates for Example 5.C in Each Iteration

Iteration	σ_1^2	σ_2^2	σ_3^2	σ_4^2
Input	—	1.00	2.00	1.00
1	4.6731	0.6819	0.6240	0.3137
2	4.3172	0.8952	0.5324	1.2520
3	4.5767	0.9754	0.4847	1.2226
⋮	⋮	⋮	⋮	⋮
8	4.6638	1.0495	0.4490	1.2060
⋮	⋮	⋮	⋮	⋮
13	4.6655	1.0505	0.4486	1.2059
14	4.6655	1.0505	0.4486	1.2059

Table 5.9   Ratio of $\hat{\sigma}_i^2 / \tilde{\sigma}_i^2$

i	Ratio
1	0.87
2	17.33
3	0.87
4	0.48

From Eq. 5.2.3, the Grubbs' estimates are

$$\tilde{\sigma}_1^2 = 5.3838 \qquad \tilde{\sigma}_2^2 = 0.0606 \qquad \tilde{\sigma}_3^2 = 0.5142 \qquad \tilde{\sigma}_4^2 = 2.5224$$

Table 5.8 gives the results of each iteration in calculating the MLE's.

The ratios of the MLE's to the Grubbs' estimates are listed in Table 5.9, again showing considerable differences.   ■

EXAMPLE 5.D

Fifteen items of unknown mass are weighed on each of five scales in an experiment to compare the scales' measurement performance. The data are given in Table 5.10.

The calculations are performed as in the previous examples. The results are displayed below. The data in Table 5.10 are converted to grams, with all results

Table 5.10   Recorded Weights of 15 Items (kg) on 5 Scales

Item	Scale 1	Scale 2	Scale 3	Scale 4	Scale 5
1	9.320	9.350	9.290	9.350	9.325
2	8.345	8.380	8.325	8.400	8.350
3	4.140	4.160	4.125	4.150	4.125
4	5.700	5.725	5.690	5.750	5.675
5	15.900	15.975	15.930	15.950	15.925
6	11.700	11.740	11.800	11.775	11.775
7	13.240	13.300	13.290	13.300	13.300
8	10.730	10.760	10.725	10.850	10.725
9	12.280	12.320	12.300	12.350	12.300
10	8.070	8.100	8.090	8.100	8.050
11	18.290	18.350	18.380	18.350	18.300
12	19.840	19.900	19.900	19.950	19.850
13	15.630	15.700	15.610	15.725	15.625
14	14.660	14.725	14.700	14.750	14.675
15	1.760	1.780	1.780	1.800	1.825

then given as grams^2. (For an interesting application of this type of exercise, see Reference 3.)

$$V_{12} = 361 \qquad V_{23} = 1349 \qquad V_{35} = 1100$$
$$V_{13} = 1621 \qquad V_{24} = 765 \qquad V_{45} = 1738$$
$$V_{14} = 953 \qquad V_{25} = 1096$$
$$V_{15} = 930 \qquad V_{34} = 1936$$

$$S_1 = 3865 \qquad S_4 = 5392$$

$$S_2 = 3571 \qquad S_5 = 4864$$

$$S_3 = 6006$$

$$V_T = 11849$$

$$\tilde{\sigma}_1^2 = 300.92 \qquad \tilde{\sigma}_4^2 = 809.92$$

$$\tilde{\sigma}_2^2 = 202.92 \qquad \tilde{\sigma}_5^2 = 633.92$$

$$\tilde{\sigma}_3^2 = 1014.58$$

Table 5.11 Parameter Estimates for Example 5.D in Each Iteration

Iteration	σ_1^2	σ_2^2	σ_3^2	σ_4^2	σ_5^2
Input	—	180	1120	800	670
1	195.05	153.21	1098.70	652.21	747.32
2	206.59	142.47	1106.39	645.60	755.72
3	209.91	140.91	1106.33	644.62	757.08
⋮	⋮	⋮	⋮	⋮	⋮
5	210.55	140.64	1106.31	644.43	757.33
⋮	⋮	⋮	⋮	⋮	⋮
7	210.57	140.63	1106.31	644.43	757.34

Table 5.12 Ratio of $\hat{\sigma}_i^2/\tilde{\sigma}_i^2$

i	Ratio
1	0.70
2	0.69
3	1.09
4	0.80
5	1.19

The MLE's are displayed in Table 5.11. The ratios of the MLE's to the Grubbs' estimates are listed in Table 5.12. ∎

EXAMPLE 5.E

In the same type of exercise as described in Example 5.D, 39 items were weighed on each of 14 scales in a round-robin experiment involving several laboratories. The data are given in Table 5.13, and V_{ij} is given in Table 5.14 in kg^2.

Without displaying the intermediate steps, the Grubbs' estimates and the MLE's are listed in Table 5.15, along with the ratio $\hat{\sigma}_i^2/\tilde{\sigma}_i^2$. Using rounded Grubbs' estimates as inputs to the program used to compute the MLE's, convergence to the nearest hundredth of a gram² occurred with only eight iterations despite the large value for N.

We have occasion later in the book to return to these examples. ∎

Table 5.13   Recorded Weights of 39 Items (kg) on 14 Scales

Items	Scales 1–14
1	61.135 61.150 61.140 61.150 61.150 61.150 61.140 61.130 61.150 61.130 61.140 61.160 61.120 61.140
2	63.875 63.880 63.880 63.880 63.890 63.880 63.870 63.870 63.890 63.860 63.880 63.900 63.860 63.880
3	60.185 60.195 60.180 60.190 60.190 60.180 60.180 60.180 60.170 60.180 60.200 60.200 60.180 60.180
4	66.760 66.755 66.750 66.760 66.780 66.770 66.760 66.750 66.740 66.750 66.760 66.760 66.740 66.760
5	70.895 70.905 70.900 70.900 70.900 70.890 70.900 70.890 70.880 70.890 70.900 70.900 70.860 70.880
6	62.140 62.155 62.150 62.160 62.160 62.150 62.140 62.140 62.150 62.140 62.160 62.160 62.120 62.140
7	64.880 64.890 64.890 64.900 64.900 64.890 64.880 64.880 64.870 64.890 64.880 64.900 64.860 64.880
8	61.190 61.200 61.190 61.200 61.200 61.190 61.190 61.190 61.180 61.190 61.200 61.220 61.180 61.180
9	67.755 67.765 67.760 67.770 67.780 67.780 67.760 67.760 67.750 67.760 67.760 67.780 67.740 67.760
10	71.900 71.900 71.910 71.900 71.910 71.900 71.900 71.900 71.890 71.900 71.900 71.900 71.880 71.900
11	66.020 66.030 66.030 66.030 66.040 66.030 66.010 66.010 66.030 66.010 66.020 66.040 66.000 66.020
12	62.335 62.340 62.330 62.330 62.340 62.340 62.320 62.330 62.320 62.330 62.340 62.340 62.320 62.340
13	68.900 68.910 68.910 68.910 68.920 68.910 68.900 68.900 68.890 68.900 68.900 68.920 68.880 68.900
14	73.035 73.040 73.050 73.050 73.050 73.040 73.040 73.030 73.020 73.040 73.040 73.040 73.020 73.040
15	65.070 65.075 65.060 65.070 65.080 65.070 65.050 65.060 65.050 65.060 65.080 65.080 65.040 65.060
16	71.635 71.640 71.640 71.650 71.660 71.650 71.630 71.630 71.620 71.640 71.640 71.640 71.620 71.640
17	75.770 75.780 75.780 75.780 75.780 75.770 75.780 75.770 75.760 75.770 75.780 75.780 75.760 75.760
18	78.655 78.660 78.650 78.650 78.670 78.680 78.640 78.640 78.640 78.640 78.640 78.640 78.620 78.640
19	60.455 60.470 60.450 60.460 60.470 60.470 60.450 60.460 60.440 60.460 60.460 60.480 60.440 60.460
20	67.025 67.040 67.030 67.040 67.050 67.040 67.020 67.030 67.020 67.040 67.020 67.040 67.000 67.020

Table 5.13   Recorded Weights of 39 Items (kg) on 14 Scales

Items	Scales 1–14
21	63.340 63.355 63.330 63.340 63.350 63.340 63.330 63.330 63.320 63.330 63.340 63.340 63.320 63.340
22	69.905 69.910 69.920 69.920 69.930 69.920 69.910 69.910 69.900 69.910 69.900 69.920 69.880 69.900
23	74.045 74.050 74.060 74.060 74.060 74.050 74.050 74.040 74.030 74.040 74.040 74.060 74.020 74.040
24	66.080 66.085 66.070 66.080 66.080 66.080 66.060 66.070 66.060 66.070 66.080 66.080 66.060 66.080
25	72.640 72.650 72.650 72.660 72.660 72.660 72.640 72.640 72.630 72.650 72.640 72.640 72.620 72.640
26	76.775 76.785 76.790 76.790 76.790 76.780 76.780 76.770 76.770 76.780 76.780 76.780 76.760 76.780
27	79.650 79.670 79.660 79.660 79.670 79.660 79.640 79.640 79.640 79.640 79.660 79.640 79.640 79.660
28	67.215 67.225 67.210 67.220 67.220 67.220 67.200 67.210 67.200 67.210 67.220 67.220 67.200 67.200
29	73.780 73.775 73.800 73.800 73.800 73.800 73.780 73.780 73.770 73.790 73.780 73.780 73.760 73.780
30	77.915 77.925 77.930 77.920 77.930 77.920 77.920 77.910 77.910 77.910 77.920 77.920 77.900 77.920
31	80.800 80.805 80.800 80.800 80.810 80.810 80.780 80.780 80.780 80.780 80.800 80.780 80.780 80.800
32	83.530 83.540 83.530 83.530 83.540 83.540 83.520 83.520 83.520 83.520 83.540 83.540 83.500 83.520
33	60.225 68.235 68.210 68.220 68.240 68.230 68.210 68.210 68.200 68.210 68.220 68.220 68.200 68.220
34	74.785 74.800 74.810 74.810 74.810 74.810 74.790 74.790 74.780 74.790 74.780 74.780 74.760 74.780
35	78.925 78.940 78.930 78.930 78.930 78.930 78.920 78.920 78.910 78.920 78.920 78.940 78.900 78.920
36	81.800 81.810 81.810 81.800 81.810 81.810 81.790 81.790 81.790 81.790 81.800 81.800 81.800 81.800
37	84.540 84.545 84.540 84.540 84.540 84.550 84.530 84.520 84.520 84.530 84.540 84.540 84.520 84.540
38	85.685 85.690 85.680 85.680 85.680 85.690 85.670 85.660 85.660 85.670 85.680 85.680 85.660 85.680
39	86.695 86.695 86.680 86.690 86.690 86.690 86.670 86.670 86.670 86.670 86.680 86.680 86.660 86.680

Table 5.14 V_{ij} for Table 5.13 Data

V	1	2 =	0.000029	V	3	9 =	0.000089	V	6	12 =	0.000211
V	1	3 =	0.000097	V	3	10 =	0.000071	V	6	13 =	0.000127
V	1	4 =	0.000072	V	3	11 =	0.000131	V	6	14 =	0.000079
V	1	5 =	0.000066	V	3	12 =	0.000196	V	7	8 =	0.000041
V	1	6 =	0.000064					V	7	9 =	0.000093
V	1	7 =	0.000080	V	3	13 =	0.000123	V	7	10 =	0.000052
V	1	8 =	0.000061	V	3	14 =	0.000099	V	7	11 =	0.000096
V	1	9 =	0.000097	V	4	5 =	0.000041	V	7	12 =	0.000122
V	1	10 =	0.000082	V	4	6 =	0.000073	V	7	13 =	0.000104
V	1	11 =	0.000049	V	4	7 =	0.000045	V	7	14 =	0.000094
V	1	12 =	0.000150	V	4	8 =	0.000039	V	8	9 =	0.000084
V	1	13 =	0.000058	V	4	9 =	0.000078	V	8	10 =	0.000022
V	1	14 =	0.000043	V	4	10 =	0.000033	V	8	11 =	0.000075
V	2	3 =	0.000110	V	4	11 =	0.000094	V	8	12 =	0.000090
V	2	4 =	0.000075	V	4	12 =	0.000134	V	8	13 =	0.000087
V	2	5 =	0.000078	V	4	13 =	0.000106	V	8	14 =	0.000082
V	2	6 =	0.000084	V	4	14 =	0.000084	V	9	10 =	0.000127
V	2	7 =	0.000094	V	5	6 =	0.000041	V	9	11 =	0.000100
V	2	8 =	0.000065	V	5	7 =	0.000068	V	9	12 =	0.000106
V	2	9 =	0.000091	V	5	8 =	0.000044	V	9	13 =	0.000117
V	2	10 =	0.000092	V	5	9 =	0.000083	V	9	14 =	0.000096
V	2	11 =	0.000055	V	5	10 =	0.000058	V	10	11 =	0.000104
V	2	12 =	0.000138	V	5	11 =	0.000106	V	10	12 =	0.000130
V	2	13 =	0.000069	V	5	12 =	0.000156	V	10	13 =	0.000109
V	2	14 =	0.000063	V	5	13 =	0.000110	V	10	14 =	0.000094
V	3	4 =	0.000039	V	5	14 =	0.000068	V	11	12 =	0.000120
V	3	5 =	0.000073	V	6	7 =	0.000112	V	11	13 =	0.000050
V	3	6 =	0.000099	V	6	8 =	0.000083	V	11	14 =	0.000067
V	3	7 =	0.000048	V	6	9 =	0.000113	V	12	13 =	0.000154
V	3	8 =	0.000071	V	6	10 =	0.000087	V	12	14 =	0.000165
				V	6	11 =	0.000139	V	13	14 =	0.000046

Table 5.15   Parameter Estimates in Grams^2

Scale	$\tilde{\sigma}_i^2$ (Grubbs)	$\hat{\sigma}_i^2$ (MLE)	$\hat{\sigma}_i^2/\tilde{\sigma}_i^2$
1	27.62	30.51	1.10
2	35.53	37.89	1.07
3	52.45	46.22	0.88
4	24.70	21.89	0.89
5	27.28	28.14	1.03
6	57.95	52.01	0.90
7	36.03	33.24	0.92
8	18.95	17.89	0.94
9	54.78	57.48	1.05
10	37.03	30.18	0.82
11	47.45	50.71	1.07
12	104.62	100.58	0.96
13	53.62	56.52	1.05
14	38.62	41.64	1.08

5.4 INFERENCE

Statistical inference problems include both the construction of interval estimates and the testing of hypotheses. In Section 5.4.1, interval estimation is discussed, while in Section 5.4.2 problems in hypothesis testing are considered.

5.4.1 Confidence Intervals on σ_i^2

Confidence Intervals Based on Grubbs' Estimates. In the case of $N = 3$ treated in Chapter 4, a confidence interval on σ_i^2 was constructed by calculating its variance given by Grubbs (1). Equations 3.4.2 and 3.4.3 were then used to find the approximate number of degrees of freedom associated with the Grubbs' estimate of σ_i^2 and to construct the approximate $100(p_2 - p_1)\%$ confidence interval. The same approach is used for $N \geq 4$.

For general N, Grubbs (1) gives the variance of $\tilde{\sigma}_i^2$ given by Eq. 5.4.1 in his Eq. 25. For simplicity in exposition, this variance is given for $i = 1$; the formulas for $i \neq 1$ may be found by rotation of the subscripts.

$$\operatorname{var} \tilde{\sigma}_1^2 = \frac{2\sigma_1^4}{n-1} + \frac{4\left[\sum\limits_{k=2}^{N} \sigma_1^2\sigma_k^2 + \sum\limits_{k=2}^{N-1}\sum\limits_{j>k} \sigma_k^2\sigma_j^2/(N-2)^2\right]}{(n-1)(N-1)^2} \tag{5.4.1}$$

Note that for $N = 3$, this reduces to Eq. 4.4.1 when $i = 1$. As before, the unknown σ_k^2's are replaced by their estimates in calculating $\operatorname{var} \tilde{\sigma}_1^2$.

Before the construction of approximate confidence intervals for the Grubbs' estimates is illustrated, the construction of limits for the MLE's is described.

Confidence Intervals Based on Large-Sample Theory. By large-sample theory (4), the MLE's $\hat{\sigma}_i^2$ are, for large n, approximately distributed by the multivariate normal distribution with means σ_i^2 and with the variance-covariance matrix being the inverse of H, where H is the symmetric matrix with element $-E[\partial^2 L/\partial(\sigma_i^2)^2]$ in position (i, i) and element $-E(\partial^2 L/\partial\sigma_i^2\,\partial\sigma_j^2)$ in position (i, j). Appendix 5.C shows that

$$\frac{\partial^2 L}{\partial(\sigma_i^2)^2} = -\frac{0.5nb_{0i}^2}{\left(1 + b_{0i}\sigma_i^2\right)^2} \tag{5.4.2}$$

and

$$\frac{\partial^2 L}{\partial\sigma_i^2\,\partial\sigma_j^2} = \frac{0.5n}{\sigma_j^4\left(1 + b_{0i}\sigma_i^2\right)^2}\left\{1 + 2b_{0i}\sigma_i^2 - \frac{n-1}{n}\left[b_{0i}V_{ij} + b_{1i} - \sum_{k\neq i}\frac{V_{jk}}{\sigma_k^2}\right]\right\}$$

$$\tag{5.4.3}$$

where b_{0i} and b_{1i} are defined in Eqs. 5.3.2 and 5.3.3, respectively. In evaluating these second order derivatives, the parameter values are replaced by their estimates.

If a joint confidence region on the parameters is desired, use is made of the quantity u defined in Eq. 5.4.4. This quantity is distributed approximately as chi-square with N degrees of freedom:

$$u = \sum_{i=1}^{N} \sum_{j=1}^{N} a_{ij} (\hat{\sigma}_i^2 - \sigma_i^2)(\hat{\sigma}_j^2 - \sigma_j^2) \tag{5.4.4}$$

where

$$a_{ii} = \frac{\partial^2 L}{\partial^2 (\sigma_i^2)^2} \tag{5.4.5}$$

and

$$a_{ij} = \frac{\partial^2 L}{\partial \sigma_i^2 \, \partial \sigma_j^2} \tag{5.4.6}$$

are evaluated using Eqs. 5.4.2 and 5.4.3. Example 4.H demonstrated the construction of a confidence region for $N = 3$, and the same procedure is followed for $N \geq 4$.

EXAMPLE 5.F

In Example 5.A, the Grubbs' estimates were found to be as follows:

$$\tilde{\sigma}_1^2 = 0.005077 \qquad \tilde{\sigma}_4^2 = 0.067171$$

$$\tilde{\sigma}_2^2 = 0.014457 \qquad \tilde{\sigma}_5^2 = 0.108759$$

$$\tilde{\sigma}_3^2 = 0.020752 \qquad \tilde{\sigma}_6^2 = 0.043269$$

For $n = 43$, approximate 95% confidence limits are found on $\sigma_1, \sigma_2, \ldots, \sigma_6$ by applying Eq. 5.4.1, with $\tilde{\sigma}_i^2$ replacing σ_i^2:

$$\text{var } \tilde{\sigma}_1^2 = \frac{(5077)^2}{21} + 4\{(5077)(14{,}457 + \cdots + 43{,}269)$$

$$+ [(1445)(20{,}752) + \cdots + (108{,}759)(43{,}269)]/16\}10^{-12}/1050$$

$$= 1.1609 \times 10^{-5}$$

$$\text{var } \tilde{\sigma}_2^2 = 2.8372 \times 10^{-5} \qquad \text{var } \tilde{\sigma}_3^2 = 4.3969 \times 10^{-5} \qquad \text{var } \tilde{\sigma}_4^2 = 2.6675 \times 10^{-4}$$

$$\text{var } \tilde{\sigma}_5^2 = 6.2757 \times 10^{-4} \qquad \text{var } \tilde{\sigma}_6^2 = 1.2833 \times 10^{-4}$$

Table 5.16 Construction of Confidence Intervals (Grubbs)

i	$\chi^2_{0.025}(df_i)/df_i$	$\chi^2_{0.975}(df_i)/df_i$	Lower Limit on σ_i^2	Upper Limit on σ_i^2	Lower Limit on σ_i	Upper Limit on σ_i
1	0.139	2.70	0.001880	0.036525	0.043	0.191
2	0.413	1.84	0.007857	0.035005	0.089	0.187
3	0.476	1.72	0.012065	0.043597	0.110	0.209
4	0.581	1.53	0.043903	0.115613	0.210	0.340
5	0.600	1.50	0.072506	0.181265	0.269	0.426
6	0.555	1.58	0.027385	0.077962	0.165	0.279

The respective approximate numbers of degrees of freedom (df) are calculated from Eq. 3.4.2:

$$df_1 = \frac{2(0.005077)^2}{1.1609 \times 10^{-5}} = 4.4$$

$$df_2 = 14.7 \qquad df_3 = 19.6 \qquad df_4 = 33.8$$

$$df_5 = 37.7 \qquad df_6 = 29.2$$

To construct approximate 95% confidence intervals on σ_i, one must first find them on σ_i^2. The $\chi^2_{0.025}(df_i)/df_i$ and $\chi^2_{0.975}(df_i)/df_i$ values are found by interpolation in Table 1.2. The two-sided inequality Eq. 3.4.3 is then solved for σ_i^2. By taking square roots, the limits on σ_i are found. The results of the calculations are given in Table 5.16.

Continuing with this example, large-sample theory for the MLE's is now used to construct the confidence intervals. In Example 4.A, the MLE's were found to be

$$\hat{\sigma}_1^2 = 0.006151 \qquad \hat{\sigma}_4^2 = 0.072838$$

$$\hat{\sigma}_2^2 = 0.008628 \qquad \hat{\sigma}_5^2 = 0.113986$$

$$\hat{\sigma}_3^2 = 0.024503 \qquad \hat{\sigma}_6^2 = 0.031918$$

The H matrix is constructed from Eqs. 5.4.2 and 5.4.3. The results are

$$H = \begin{pmatrix} 180{,}930 & 60{,}430 & 4{,}296 & -1{,}405 & -443 & 12{,}075 \\ & 137{,}248 & 5{,}718 & 2{,}290 & 1{,}133 & -8{,}590 \\ & & 28{,}405 & -99 & -148 & 1{,}057 \\ & \text{(the matrix is symmetric)} & & 3{,}760 & -36 & 182 \\ & & & & 1{,}578 & -80 \\ & & & & & 17{,}709 \end{pmatrix}$$

Table 5.17　Construction of Confidence Intervals (MLE's)

i	$\chi^2_{0.025}(df_i)/df_i$	$\chi^2_{0.975}(df_i)/df_i$	Lower Limit on σ_i^2	Upper Limit on σ_i^2	Lower Limit on σ_i	Upper Limit on σ_i
1	0.334	2.03	0.003030	0.018416	0.055	0.136
2	0.424	1.82	0.004741	0.020349	0.069	0.143
3	0.580	1.53	0.016015	0.042247	0.127	0.206
4	0.605	1.49	0.048885	0.120393	0.221	0.347
5	0.613	1.48	0.077018	0.185948	0.278	0.431
6	0.568	1.56	0.020460	0.056194	0.143	0.237

The diagonal elements in H^{-1} provide the approximate variances of $\hat{\sigma}_i^2$. The results are

$$\text{var}\,\hat{\sigma}_1^2 = 7.2944 \times 10^{-6} \qquad \text{var}\,\hat{\sigma}_4^2 = 2.7339 \times 10^{-4}$$

$$\text{var}\,\hat{\sigma}_2^2 = 9.6745 \times 10^{-6} \qquad \text{var}\,\hat{\sigma}_5^2 = 6.4119 \times 10^{-4}$$

$$\text{var}\,\hat{\sigma}_3^2 = 3.5699 \times 10^{-5} \qquad \text{var}\,\hat{\sigma}_6^2 = 6.4981 \times 10^{-5}$$

From Eq. 3.4.2, the approximate numbers of degrees of freedom are

$$df_1 = 10.4 \qquad df_4 = 38.8$$

$$df_2 = 15.4 \qquad df_5 = 40.5$$

$$df_3 = 33.6 \qquad df_6 = 31.4$$

Note how these respective numbers of degrees of freedom compare with those calculated for the Grubbs' estimates.

Table 5.18　Confidence Interval Widths: (Upper Limit/Lower Limit) on σ_i

i	Grubbs	MLE
1	4.44	2.47
2	2.10	2.07
3	1.90	1.62
4	1.62	1.57
5	1.58	1.55
6	1.69	1.66

Table 5.17 summarizes the calculations leading to the respective approximate 95% confidence limits derived from the MLE's.

In closing the example, it is interesting to compare the widths of the confidence intervals based on the Grubbs' estimates and on the MLE's. Table 5.18 gives the ratio (upper limit on σ_i)/(lower limit on σ_i) calculated from the last columns of Tables 5.16 and 5.17.   ∎

5.4.2 Hypothesis Testing

As mentioned in Section 4.4.2 for $N = 3$, the hypotheses tests considered for $N \geq 4$ are large-sample likelihood-ratio tests related to those formulated by Jaech (5) for $N \geq 3$ methods. In the original formulation of these large-sample tests, Grubbs' estimates formed the basis for the test statistics rather than the then-as-yet-undeveloped MLE's. For $N = 3$, this made no difference in the applications of the tests, since the MLE's are related to the Grubbs' estimates within the factor $(n - 1)/n$. This is not the case for $N \geq 4$ where, as we have seen, the Grubbs' estimates and the MLE's may differ appreciably from one another. Thus the discussion of the hypothesis-testing methods that follows is restricted to MLE's.

The logarithm of the likelihood, denoted by L, is given in Appendix 5.B as Eq. 5.B.1. It is repeated here for reference.

$$L = C - 0.5n \ln Q - 0.5(n - 1)(V/Q) \sum_{i=1}^{N-1} \sum_{j>i} \frac{V_{ij}}{\sigma_i^2 \sigma_j^2} \quad (5.4.7)$$

where

$$V = \sigma_1^2 \sigma_2^2 \cdots \sigma_N^2 \quad (5.4.8)$$

and

$$Q = V \sum_{i=1}^{N} \frac{1}{\sigma_i^2} \quad (5.4.9)$$

As a large-sample test of any given hypothesis about the σ_i^2's, the maximum value of L is found under that hypothesis and denoted by $L(\hat{\omega})$. The value of L corresponding to the MLE's of the σ_i^2's is also calculated. This is denoted by $L(\hat{\Omega})$, and the $N \geq 4$ equivalent of $L(\hat{\Omega})$ for $N = 3$ (Eq. 4.4.5) is

$$L(\hat{\Omega}) = C - 0.5n \ln \hat{Q} - 0.5n(N - 1) \quad (5.4.10)$$

where $\hat{Q}$ is given by Eq. 5.4.9 with the σ_i^2's replaced by their MLE's.

Having found $L(\hat{\Omega})$ and $L(\hat{\omega})$ for a specified hypothesis, large-sample theory tells us that, under this specified hypothesis,

$$\lambda = -2[L(\hat{\omega}) - L(\hat{\Omega})] \tag{5.4.11}$$

is distributed approximately as chi-square with $(N - r)$ degrees of freedom, where r is the number of parameters estimated under the hypothesis (not including the α_i's, which are calculated in both the ω and Ω spaces). Thus the test statistic is λ, and the specified hypothesis is rejected if λ exceeds the prespecified critical value that is a function of the $(N - r)$ degrees of freedom and the value chosen for α, the type 1 error probability.

In calculating $L(\hat{\omega})$ for a specified hypothesis, one hypothesis of particular interest is that the σ_i^2's are all equal. Letting σ_0^2 denote this common value such that the hypothesis is written

$$H_0: \quad \sigma_1^2 = \sigma_2^2 = \cdots = \sigma_N^2 = \sigma_0^2 \tag{5.4.12}$$

it is shown in Appendix 5.E that the MLE of σ_0^2 is

$$\hat{\sigma}_0^2 = \frac{(n-1)\sum\limits_{i=1}^{N-1}\sum\limits_{j>i} V_{ij}}{nN(N-1)} \tag{5.4.13}$$

and that the corresponding likelihood is

$$L(\hat{\omega}) = C - 0.5n\left[\ln N + (N-1)\ln\hat{\sigma}_0^2 + (N-1)\right] \tag{5.4.14}$$

The test of the hypothesis specified in Eq. 5.4.12 is illustrated in the examples to follow, as are tests of other hypotheses of potential interest.

EXAMPLE 5.G

In Example 5.A dealing with measurements of reactor fuel pellet densities, the Grubbs' estimates and the MLE's were found to be as follows, in $(\%TD)^2$. Recall that $n = 43$ for these data.

Grubbs	MLE's
$\tilde{\sigma}_1^2 = 0.005077$	$\hat{\sigma}_1^2 = 0.006151$
$\tilde{\sigma}_2^2 = 0.014457$	$\hat{\sigma}_2^2 = 0.008628$
$\tilde{\sigma}_3^2 = 0.020752$	$\hat{\sigma}_3^2 = 0.024503$
$\tilde{\sigma}_4^2 = 0.067171$	$\hat{\sigma}_4^2 = 0.072838$
$\tilde{\sigma}_5^2 = 0.108759$	$\hat{\sigma}_5^2 = 0.113986$
$\tilde{\sigma}_6^2 = 0.043269$	$\hat{\sigma}_6^2 = 0.031918$

Test the hypothesis: H_0: $\sigma_1^2 = \sigma_2^2 = \cdots = \sigma_6^2 = \sigma_0^2$ with the significance level of $\alpha = 0.05$.

The large-sample test discussed in Section 5.4.2 is based on the MLE's. The Grubbs' estimates are given above only for comparison, since it is of interest to compute the likelihood corresponding to the Grubbs' estimates and compare it with $L(\hat{\Omega})$ of Eq. 5.4.10, that is, the likelihood for the MLE's. Before doing this, the above hypothesis is tested.

To calculate $L(\hat{\Omega})$ from Eq. 5.4.10, $\hat{Q}$ is found first from Eqs. 5.4.8 and 5.4.9 with σ_i^2 replaced by $\hat{\sigma}_i^2$ for all i. We have

$$\hat{V} = 3.4460 \times 10^{-10}$$

$$\hat{Q} = 1.2858 \times 10^{-7}$$

Therefore, from Eq. 5.4.10,

$$L(\hat{\Omega}) = C - 21.5\ln(1.2858 \times 10^{-7}) - 107.5$$

$$= C + 233.634$$

To calculate $L(\hat{\omega})$ from Eq. 5.4.14, $\hat{\sigma}_0^2$ is calculated from Eq. 5.4.13 where, from the example cited earlier, the sum of the V_{ij} values was found to be 1.297420. Then,

$$\hat{\sigma}_0^2 = 0.042242$$

and from Eq. 5.4.14,

$$L(\hat{\omega}) = C - 21.5(\ln 6 + 5\ln 0.042242 + 5)$$

$$= C + 194.145$$

λ is calculated from Eq. 5.4.11:

$$\lambda = -2(194.145 - 233.634)$$

$$= 78.98$$

The number of degrees of freedom is $(6 - 1) = 5$. For $\alpha = 0.05$, the test critical value is 11.07. Since λ exceeds the critical value, the hypothesis of equal precisions is rejected.

As a matter of interest, $L(\tilde{\Omega})$ is also calculated based on the Grubbs' estimates rather than the MLE's. Equation 5.4.7 is evaluated with σ_i^2 replaced by $\tilde{\sigma}_i^2$ for all i. From Eqs. 5.4.8 and 5.4.9,

$$\tilde{V} = 4.8147 \times 10^{-10}$$

$$\tilde{Q} = 1.7406 \times 10^{-7}$$

Then,

$$L(\tilde{\Omega}) = C - 21.5 \ln(1.7406 \times 10^{-7})$$

$$- (21)(0.002766)\left[\frac{0.015412}{(0.005077)(0.014457)} \right.$$

$$\left. + \cdots + \frac{0.172345}{(0.108759)(0.043269)} \right]$$

$$= C + 334.623 - 102.766$$

$$= C + 231.857$$

Note how this compares with the corresponding value of $C + 233.634$ computed using the MLE's. Of course, the likelihood for the MLE's will always be larger by definition.   ■

EXAMPLE 5.H

For the data of Example 5.A, test the hypothesis

$$H_{01}: \quad \sigma_1^2 = \sigma_2^2 = \sigma_3^2 = \sigma_0^2$$

$$\sigma_4^2 = \sigma_5^2 = \sigma_6^2 = 4\sigma_0^2$$

against the alternative that at least one of the inequalities does not hold. This is a reasonable hypothesis to test, since Methods 1, 2, and 3 were geometric measurement methods while Methods 4, 5, and 6 were immersion methods. Before the measurement experiment was conducted, the standard deviation of the immersion method was hypothesized to be twice that of the geometric method.

First, $L(\hat{\Omega})$ is the same as in Example 5.G.

$$L(\hat{\Omega}) = C + 233.634$$

To compute $L(\hat{\omega})$ corresponding to H_{01}, the MLE of σ_0^2 under H_{01} must be found. The expression for L in Eq. 5.4.7 is rewritten to replace σ_i^2, $i = 1, 2, \ldots, 6$, by σ_0^2 for $i = 1, 2, 3$ and by $4\sigma_0^2$ for $i = 4, 5, 6$. First, V and Q

become, respectively,

$$V = 64\sigma_0^{12}$$

$$Q = 64\sigma_0^{12}\left(\frac{3}{\sigma_0^2} + \frac{3}{4\sigma_0^2}\right)$$

$$= 240\sigma_0^{10}$$

Then,

$$L = C - 0.5n \ln 240 - 2.5n \ln \sigma_0^2$$

$$- 2(n-1)\left[(V_{12} + V_{13} + V_{23}) + (V_{14} + V_{15} + V_{16} + V_{24} + V_{25}\right.$$

$$+ V_{26} + V_{34} + V_{35} + V_{36})/4 + (V_{45} + V_{46} + V_{56})/16\left]\right./15\sigma_0^2$$

$$\frac{\partial L}{\partial \sigma_0^2} = \frac{-2.5n}{\sigma_0^2} + \frac{2(n-1)}{15\sigma_0^4}[\cdots] = 0$$

where $[\cdots]$ denotes the indicated sum in the square brackets, which is calculated to be 0.303107. Thus the MLE of σ_0^2 under this hypothesis is, for $n = 43$,

$$\hat{\sigma}_0^2 = \frac{(84)(0.303107)}{(37.5)(43)} = 0.015790$$

In evaluating the expression for L, note that

$$\frac{-84[\cdots](37.5)(43)}{(15)(84)[\cdots]} = -107.5$$

or, more generally, $-0.5n(N-1)$, the same as the last term in Eq. 5.4.10 and in Eq. 5.4.14. Thus very simply,

$$L(\hat{\omega}) = C - 21.5 \ln 240 - (2.5)(43)\ln 0.015790 - 107.5$$

$$= C + 220.617$$

Then, from Eq. 5.4.11,

$$\lambda = -2(220.617 - 233.634)$$

$$= 26.03$$

The number of degrees of freedom is again $(6 - 1) = 5$. For $\alpha = 0.05$, the critical value of 11.07 is still exceeded. Thus at least one of the equality signs in the stated hypothesis is not supported by the data.   ∎

EXAMPLE 5.I

In Example 5.C, data were given for measurements of uranium concentration made on each of eight samples by each of four laboratories. Tabulated below are the calculated V_{ij} values and the MLE's of the parameters from Example 5.C.

$$V_{12} = 4.021455 \qquad V_{23} = 1.926619$$

$$V_{13} = 5.969130 \qquad V_{24} = 2.654039$$

$$V_{14} = 9.258080 \qquad V_{34} = 1.613654$$

$$\hat{\sigma}_1^2 = 4.6655$$

$$\hat{\sigma}_2^2 = 1.0505$$

$$\hat{\sigma}_3^2 = 0.4486$$

$$\hat{\sigma}_4^2 = 1.2059$$

Although the sample size (eight) is too small to permit valid application of large-sample theory, this example is included here because it illustrates how poor the Grubbs' estimates may be in a statistical sense. The likelihoods are correctly calculated by the methods presented in this chapter, but the distribution of λ will not be closely approximated by the chi-square unless n is sufficiently large.

With this note of caution, we proceed to test the hypothesis that $\sigma_1^2 = \sigma_2^2 = \sigma_3^2 = \sigma_4^2 = \sigma_0^2$ for $\alpha = 0.05$. First, $\hat{Q}$ is found from Eqs. 5.4.8 and 5.4.9 with σ_i^2 replaced by $\hat{\sigma}_i^2$ for all i.

$$\hat{V} = (4.6655) \cdots (1.2059) = 2.6513$$

$$\hat{Q} = 2.6513\left(\frac{1}{4.6655} + \cdots + \frac{1}{1.2059}\right) = 11.2009$$

From Eq. 5.4.10,

$$L(\hat{\Omega}) = C - 4\ln 11.2009 - 12$$

$$= C - 21.664$$

Under the hypothesis, σ_0^2 is calculated from Eq. 5.4.13, where the V_{ij}'s sum to 25.442977.

$$\sigma_0^2 = \frac{(7)(25.442977)}{(8)(4)(3)} = 1.8552$$

The corresponding likelihood is

$$L(\hat{\omega}) = C - 4(\ln 4 + 3\ln 1.8552 + 3)$$

$$= C - 24.961$$

λ is calculated from Eq. 5.4.11:

$$\lambda = -2(-24.961 + 21.664) = 6.594$$

For $\alpha = 0.05$, and for $(4 - 1) = 3$ degrees of freedom, the critical value of 7.81 is not exceeded by λ, and so the hypothesis is not rejected.

We now calculate the likelihood corresponding to the Grubbs' estimates. These were found in Example 5.C to be

$$\tilde{\sigma}_1^2 = 5.3838$$

$$\tilde{\sigma}_2^2 = 0.0606$$

$$\tilde{\sigma}_3^2 = 0.5142$$

$$\tilde{\sigma}_4^2 = 2.5224$$

The likelihood given by Eq. 5.4.7 is evaluated with σ_i^2 replaced by $\tilde{\sigma}_i^2$ for all i. From Eqs. 5.4.8 and 5.4.9,

$$\tilde{V} = 0.4232$$

$$\tilde{Q} = 19.0286\tilde{V} = 8.0529$$

Then,

$$L(\hat{\Omega}) = C - 4\ln 8.0529 - 3.5(19.0286)^{-1}$$

$$\times \left[\frac{4.021455}{(5.3838)(0.0606)} + \cdots + \frac{1.613654}{(0.5142)(2.5224)} \right]$$

$$= C - 8.344 - 17.584 = C - 25.928$$

Note that in this example, L is actually *smaller* than $L(\hat{\omega})$ corresponding to the hypothesis of equal precisions. Thus in this example, the Grubbs' estimates are poor in a statistical sense, that is, using the likelihood value as a criterion of quality. ∎

EXAMPLE 5.J

In Example 5.G, test the following hypothesis at the $\alpha = 0.05$ level:

$$H_{02}: \quad \sigma_1^2 = 0$$

In practical terms, this would be interpreted to mean that the measurement error for Method 1 is negligibly small and may be assigned the value of zero. Again, as in Examples 5.G and 5.H,

$$L(\hat{\Omega}) = C + 233.634$$

Under H_{02}, the MLE's of σ_i^2, $i \neq 1$, corresponding to $\sigma_1^2 = 0$ are given by Eq. 5.5.1. For these data:

$$\hat{\sigma}_2^2 = \frac{(42)(0.015412)}{43} = 0.015054$$

$$\hat{\sigma}_3^2 = \frac{(42)(0.030374)}{43} = 0.029668$$

$$\hat{\sigma}_4^2 = 0.071415$$

$$\hat{\sigma}_5^2 = 0.113681$$

$$\hat{\sigma}_6^2 = 0.043467$$

Under H_{02}, the likelihood is given by Eq. 5.D.1, where the last term is $-0.5n(N-1)$ by virtue of Eq. 5.D.4:

$$L(\hat{\omega}) = C - 21.5(-15.663157) - 107.5$$

$$= C + 229.258$$

λ is calculated from Eq. 5.4.11:

$$\lambda = -2(229.258 - 233.634)$$

$$= 8.75$$

The number of degrees of freedom is $(6 - 5) = 1$. For $\alpha = 0.05$, the test critical value is 3.84. Since λ exceeds 3.84, the hypothesis that $\sigma_1^2 = 0$ is rejected.
∎

5.5 CONSTRAINED MAXIMUM LIKELIHOOD ESTIMATION (CMLE)

In the iterative estimation process, one estimate, say that of σ_i^2, may be negative. If this occurs on the first iteration, the iterative process should be continued, for the input values may have been chosen so large as to make the estimate of the remaining parameter negative on the first iteration. Assuming, however, that the estimates converge to their final value, and that σ_i^2 is negative, then its estimate is zero and the CMLE of σ_j^2 is derived in Appendix 5.D.

$$\hat{\sigma}_j^2 = \frac{(n-1)V_{ij}}{n}; \qquad j \neq i \qquad (5.5.1)$$

Recall that this is the same estimate as for the $N = 3$ measurement methods, Eq. 4.5.1. Equation 5.5.1 has already been applied in Example 5.I, where the hypothesis was tested that $\sigma_1^2 = 0$.

If there is any doubt as to which of two parameters, σ_i^2 or σ_j^2, is closer to zero and would be estimated to be zero, then it is also shown in Appendix 5.D that $\hat{\sigma}_i^2$ is zero if

$$\prod_{k \neq i} \hat{\sigma}_k^2 < \prod_{k \neq j} \hat{\sigma}_k^2 \qquad (5.5.2)$$

and that $\hat{\sigma}_j^2$ is zero if the inequality is reversed.

Note that Eq. 5.5.1 has intuitive appeal, for if method i were, for example, assigned standard values known to have negligible error, then clearly $\sigma_i^2 = 0$, and the estimate of σ_j^2 would be given by Eq. 5.5.1.

EXAMPLE 5.K

In Example 5.D, 15 items were weighed on each of 5 scales. The data were given in Table 5.10. A sixth scale is now included in the database with the recorded item weights for this scale given in Table 5.19.

To the V_{ij} values given following Table 5.10 for the five scales, we include V_{i6} for $i = 1, 2, \ldots, 5$. As before, the variances are in grams^2.

$$V_{16} = 18 \qquad V_{46} = 958$$

$$V_{26} = 293 \qquad V_{56} = 924$$

$$V_{36} = 1596$$

Table 5.19 Item Weights for Scale 6 (kg)

Item	Scale 6	Item	Scale 6	Item	Scale 6
1	9.325	6	11.710	11	18.300
2	8.355	7	13.260	12	19.850
3	4.155	8	10.742	13	15.645
4	5.710	9	12.294	14	14.678
5	15.920	10	8.085	15	1.770

Table 5.20 Parameter Estimates for Example 5.K in Each Iteration

Iteration	σ_1^2	σ_2^2	σ_3^2	σ_4^2	σ_5^2	σ_6^2
Input	—	180	1120	800	670	170
1	62.99	235.48	1306.21	771.75	778.17	−11.11
2	29.50	293.70	1599.22	977.67	924.92	−9.14
3	26.39	282.04	1566.07	959.27	911.26	−7.21
⋮	⋮	⋮	⋮	⋮	⋮	⋮
7	20.66	273.43	1514.60	916.05	878.52	−3.38
⋮	⋮	⋮	⋮	⋮	⋮	⋮
12	19.16	272.92	1503.54	906.31	871.41	−2.30
⋮	⋮	⋮	⋮	⋮	⋮	⋮
18	18.91	272.92	1501.78	904.73	870.27	−2.11
⋮	⋮	⋮	⋮	⋮	⋮	⋮
23	18.88	272.92	1501.60	904.56	870.15	−2.09
24	18.88	272.92	1501.59	904.55	870.14	−2.09

Table 5.20 displays the MLE's for several iterations. Since $\hat{\sigma}_6^2$ is negative, it should be assigned the value of zero. From Eq. 5.5.1, the CMLE's of σ_i^2 for $i = 1, 2$, are

$$\hat{\sigma}_1^2 = \frac{14(18)}{15} = 16.80 \text{ grams}^2$$

$$\hat{\sigma}_2^2 = \frac{14(293)}{15} = 273.47 \text{ grams}^2$$

$$\hat{\sigma}_3^2 = 1489.60 \text{ grams}^2 \qquad \hat{\sigma}_4^2 = 894.13 \text{ grams}^2 \qquad \hat{\sigma}_5^2 = 862.40 \text{ grams}^2$$

Note that $\hat{\sigma}_1^2$ is also small. If σ_1^2 is set equal to zero, then

$$\hat{\sigma}_2^2 = \frac{14(361)}{15} = 336.93 \text{ grams}^2$$

$$\hat{\sigma}_3^2 = \frac{14(1621)}{15} = 1512.93 \text{ grams}^2$$

$$\hat{\sigma}_4^2 = 889.47 \text{ grams}^2 \qquad \hat{\sigma}_5^2 = 868.00 \text{ grams}^2 \qquad \hat{\sigma}_6^2 = 16.80 \text{ grams}^2$$

To see which set of estimates is preferred from the standpoint of likelihood, Eq. 5.5.2 is applied. If $\sigma_6^2 = 0$,

$$\prod_{k \neq 6} \hat{\sigma}_k^2 = 5.277 \times 10^{12}$$

If $\sigma_1^2 = 0$,

$$\prod_{k \neq 1} \hat{\sigma}_k^2 = 6.612 \times 10^{12}$$

By the inequality Eq. 5.5.2, $\hat{\sigma}_6^2 = 0$ and $\hat{\sigma}_1^2 = 16.80 \text{ grams}^2$. ∎

5.6 ESTIMATION OF RELATIVE BIASES

The discussion of this subject in Section 4.7 for $N = 3$ is easily extended to $N \geq 4$. The generalizations of Eqs. 4.7.2, 4.7.4, and 4.7.5 to N are apparent. The general form of Eq. 4.7.6 is

$$s_{\bar{x}}^2 = \frac{\sum\limits_{i=1}^{N-1} \sum\limits_{j>i} \bar{y}_{ij}^2}{N(N-1)} \tag{5.6.1}$$

EXAMPLE 5.L

In Example 5.B, dealing with the measurement of stress at 600% elongation of 7 rubber samples, for the 5 laboratories the mean differences are as follows:

$$\bar{y}_{35} = 0.1614 \qquad \bar{y}_{58} = 0.2817$$

$$\bar{y}_{36} = -0.1139 \qquad \bar{y}_{5,12} = 0.2270$$

$$\bar{y}_{38} = 0.1203 \qquad \bar{y}_{68} = -0.2341$$

$$\bar{y}_{3,12} = 0.0656 \qquad \bar{y}_{6,12} = -0.1794$$

$$\bar{y}_{56} = 0.0476 \qquad \bar{y}_{8,12} = 0.0547$$

Use these data to estimate σ_α^2, the variance between laboratories. From Eq. 5.6.1,

$$s_{\bar{x}}^2 = \frac{(0.1614)^2 + (-0.1139)^2 + \cdots + (0.0547)^2}{20}$$

$$= 0.014046$$

To apply Eq. 4.7.5, the MLE's of the σ_i^2's are, from Example 5.B,

$$\hat{\sigma}_3^2 = 0.000477 \qquad \hat{\sigma}_8^2 = 0.004386$$

$$\hat{\sigma}_5^2 = 0.002146 \qquad \hat{\sigma}_{12}^2 = 0.000531$$

$$\hat{\sigma}_6^2 = 0.003442$$

These sum to 0.010982. Therefore, the estimate of σ_α^2 is

$$\hat{\sigma}_\alpha^2 = 0.014046 - \frac{0.010982}{(7)(5)}$$

$$= 0.013732$$

and the relative $\hat{\sigma}_\alpha = [\exp(0.013732) - 1]^{0.5} = 0.118$ or 11.8% relative. (See Example 4.D.) ■

DERIVATION OF GRUBBS' ESTIMATORS

The problem is to choose σ_i^2, $i = 1, 2, \ldots, N$, to minimize

$$S = \sum_{i=1}^{N-1} \sum_{j>i} \left(V_{ij} - \sigma_i^2 - \sigma_j^2 \right)^2 \qquad (5.A.1)$$

We have

$$\frac{\partial S}{\partial \sigma_i^2} = 2\sum_{j \neq i} \left(V_{ij} - \sigma_i^2 - \sigma_j^2 \right)(-1) = 0 \qquad (5.A.2)$$

from which

$$\sum_{j \neq i} V_{ij} - (N - 1)\sigma_i^2 - \sum_{j \neq i} \sigma_j^2 = 0 \qquad (5.A.3)$$

Letting

$$S_i = \sum_{j \neq i} V_{ij} \qquad (5.A.4)$$

the N Eqs. 5.A.3 may be written

$$
\begin{aligned}
(N - 1)\sigma_1^2 + \sigma_2^2 + \sigma_3^2 + \cdots + \sigma_N^2 &= S_1 \\
\sigma_1^2 + (N - 1)\sigma_2^2 + \sigma_3^2 + \cdots + \sigma_N^2 &= S_2 \\
&\vdots \\
\sigma_1^2 + \sigma_2^2 + \cdots + (N - 1)\sigma_N^2 &= S_N
\end{aligned}
\qquad (5.A.5)
$$

Equations 5.A.5 are written in matrix notation:

$$
\begin{pmatrix}
(N - 1) & 1 & 1 & \cdots & 1 \\
1 & (N - 1) & 1 & \cdots & 1 \\
\vdots & \vdots & \vdots & \vdots & \\
1 & 1 & \cdots & 1 & (N - 1)
\end{pmatrix}
\begin{pmatrix}
\sigma_1^2 \\
\sigma_2^2 \\
\vdots \\
\sigma_N^2
\end{pmatrix}
=
\begin{pmatrix}
S_1 \\
S_2 \\
\vdots \\
S_N
\end{pmatrix}
$$

$$(5.A.6)$$

The inverse of the $N \times N$ matrix is found to be

$$
\frac{1}{2(N-1)(N-2)}
\begin{pmatrix}
(2N-3) & -1 & -1 & \cdots & -1 \\
-1 & (2N-3) & -1 & \cdots & -1 \\
\vdots & & & & \\
-1 & -1 & & \cdots & (2N-3)
\end{pmatrix}
$$

$$(5.A.7)$$

Premultiplying both sides of Eq. 5.A.6 by this inverse matrix gives the solution to Eqs. 5.A.5. The ith element in the resulting column matrix is denoted by $\tilde{\sigma}_i^2$ and is

$$
\tilde{\sigma}_i^2 = \frac{(2N-3)S_i - \sum_{j \neq i} S_j}{2(N-1)(N-2)}
\tag{5.A.8}
$$

Now, since

$$
\sum_{j=1}^{N} S_j = 2V_T
\tag{5.A.9}
$$

where

$$
V_T = \sum_{i=1}^{N-1} \sum_{j>i} V_{ij}
\tag{5.A.10}
$$

(i.e., V_T is the sum of all the V_{ij}'s), it follows that the numerator of Eq. 5.A.8 may be written

$$
(2N - 3 + 1)S_i - 2V_T = 2[(N-1)S_i - V_T]
\tag{5.A.11}
$$

Therefore,

$$
\tilde{\sigma}_i^2 = \frac{(N-1)S_i - V_T}{(N-1)(N-2)}
\tag{5.A.12}
$$

This is Eq. 5.2.3, thus completing the derivation.

DERIVATION OF MAXIMUM LIKELIHOOD ESTIMATORS (MLE'S)

The derivation of the MLE of σ_i^2 given by Eq. 5.3.1 parallels the derivation for $N = 3$ given in Appendix 4.A. The general $N \geq 4$ version of the logarithm of the likelihood corresponding to Eq. 4.A.12 for $N = 3$ is

$$L = C - 0.5n \ln Q - 0.5(n-1)(V/Q) \sum_{i=1}^{N-1} \sum_{j>i} \frac{V_{ij}}{\sigma_i^2 \sigma_j^2} \qquad (5.B.1)$$

where

$$V = \sigma_1^2 \sigma_2^2 \cdots \sigma_N^2 \qquad (5.B.2)$$

and

$$Q = V \sum_{i=1}^{N} \frac{1}{\sigma_i^2} \qquad (5.B.3)$$

so that

$$\frac{V}{Q} = \left(\sum_{i=1}^{N} \frac{1}{\sigma_i^2} \right)^{-1} \qquad (5.B.4)$$

Note that for $N = 3$, Eq. 5.B.1 reduces to Eq. 4.A.12, and Eq 5.B.3 reduces to Eq. 4.A.5.

To continue, the expression for L in Eq. 5.B.1 is rewritten to isolate terms in σ_i^2.

$$L = C - 0.5n \ln V - 0.5n \ln\left(\frac{1}{\sigma_1^2} + \frac{1}{\sigma_2^2} + \cdots + \frac{1}{\sigma_N^2} \right)$$

$$- \frac{0.5(n-1)}{1/\sigma_1^2 + 1/\sigma_2^2 + \cdots + 1/\sigma_N^2} \sum_{i=1}^{N-1} \sum_{j>i} \frac{V_{ij}}{\sigma_i^2 \sigma_j^2}$$

$$= C - 0.5n \ln \sigma_i^2 - 0.5n \sum_{j \neq i} \ln \sigma_j^2 - 0.5n \ln\left(1/\sigma_i^2 + b_{0i} \right)$$

$$- \frac{0.5(n-1)}{1/\sigma_i^2 + b_{0i}} \left(\frac{b_{1i}}{\sigma_i^2} + b_{2i} \right) \qquad (5.B.5)$$

where

$$b_{0i} = \sum_{j \neq i} \frac{1}{\sigma_j^2}$$

(5.B.6)

$$b_{1i} = \sum_{j \neq i} \frac{V_{ij}}{\sigma_j^2}$$

(5.B.7)

$$b_{2i} = \sum_{j} \sum_{\substack{k > j \\ j, k \neq i}} \frac{V_{jk}}{\sigma_j^2 \sigma_k^2}$$

(5.B.8)

Equation 5.B.5 is further reduced upon writing

$$\left(\frac{1}{\sigma_i^2} + b_{0i} \right) = \frac{1 + b_{0i}\sigma_i^2}{\sigma_i^2}$$

(5.B.9)

Then,

$$L = C - 0.5n \sum_{j \neq i} \ln \sigma_j^2 - 0.5n \ln\left(1 + b_{0i}\sigma_i^2\right)$$

$$- \frac{0.5(n-1)\left(b_{1i} + b_{2i}\sigma_i^2\right)}{1 + b_{0i}\sigma_i^2}$$

(5.B.10)

The MLE of σ_i^2 is then found by equating $\partial L/\partial \sigma_i^2$ to 0 and solving for σ_i^2.

$$\frac{\partial L}{\partial \sigma_i^2} = -\frac{0.5nb_{0i}}{1 + b_{0i}\sigma_i^2} - \frac{0.5(n-1)\left(b_{2i} - b_{0i}b_{1i}\right)}{\left(1 + b_{0i}\sigma_i^2\right)^2} = 0$$

(5.B.11)

from which

$$nb_{0i}\left(1 + b_{0i}\sigma_i^2\right) + (n-1)\left(b_{2i} - b_{0i}b_{1i}\right) = 0$$

and

$$\hat{\sigma}_i^2 = \frac{(n-1)\left(b_{0i}b_{1i} - b_{2i}\right)}{nb_{0i}^2} - \frac{1}{b_{0i}}$$

(5.B.12)

which is Eq. 5.3.1, thus completing the derivation.

DERIVATION OF CONFIDENCE INTERVALS BASED ON LARGE-SAMPLE THEORY

From Eq. 5.B.11, upon replacing σ_i^2 by v_i for simplicity in notation,

$$\frac{\partial L}{\partial v_i} = -\frac{0.5nb_{0i}}{1 + b_{0i}v_i} - \frac{0.5(n-1)(b_{2i} - b_{0i}b_{1i})}{(1 + b_{0i}v_i)^2} \qquad (5.C.1)$$

where b_{0i}, b_{1i}, and b_{2i} are defined by Eqs. 5.B.6 to 5.B.8.
 Noting that

$$\frac{\partial b_{0i}}{\partial v_i} = \frac{\partial b_{1i}}{\partial v_i} = \frac{\partial b_{2i}}{\partial v_i} = 0 \qquad (5.C.2)$$

then

$$\frac{\partial^2 L}{\partial v_i^2} = \frac{0.5nb_{0i}^2}{(1 + b_{0i}v_i)^2} + \frac{(n-1)b_{0i}(b_{2i} - b_{0i}b_{1i})}{(1 + b_{0i}v_i)^3} \qquad (5.C.3)$$

Now, since $\partial L/\partial v_i = 0$, from Eq. 5.C.1,

$$\frac{(n-1)b_{0i}(b_{2i} - b_{0i}b_{1i})}{(1 + b_{0i}v_i)^3} = -\frac{nb_{0i}^2}{(1 + b_{0i}v_i)^2} \qquad (5.C.4)$$

and hence,

$$\frac{\partial^2 L}{\partial v_i^2} = -\frac{0.5nb_{0i}^2}{(1 + b_{0i}v_i)^2} \qquad (5.C.5)$$

Thus, Eq. 5.4.2 has been derived. Next, Eq. 5.4.3 is derived. Note that

$$\frac{\partial b_{0i}}{\partial v_j} = -\frac{1}{v_j^2} \tag{5.C.6}$$

$$\frac{\partial b_{1i}}{\partial v_j} = -\frac{V_{ij}}{v_j^2} \tag{5.C.7}$$

$$\frac{\partial b_{2i}}{\partial v_j} = -\sum_{k \neq i} \frac{V_{jk}}{v_j^2 v_k} \tag{5.C.8}$$

Then,

$$\frac{\partial^2 L}{\partial v_i\, \partial v_j} = \frac{0.5n}{v_j^2(1 + b_{0i}v_i)^2} - \frac{0.5(n-1)A}{v_j^2(1 + b_{0i}v_i)^3} \tag{5.C.9}$$

where

$$A = 2v_i(b_{2i} - b_{0i}b_{1i}) + (1 + b_{0i}v_i)\left(b_{0i}V_{ij} + b_{1i} - \sum_{k \neq i} \frac{V_{jk}}{v_k}\right) \tag{5.C.10}$$

The first set of terms in A is replaced by their equivalent expression using Eq. 5.C.1,

$$2v_i(b_{2i} - b_{0i}b_{1i}) = \frac{-2nb_{0i}v_i(1 + b_{0i}v_i)}{n-1} \tag{5.C.11}$$

and hence, from Eqs. 5.C.9 and 5.C.10,

$$\frac{\partial^2 L}{\partial v_i\, \partial v_j} = \frac{0.5n}{v_j^2(1 + b_{0i}v_i)^2}\left\{1 + 2b_{0i}v_i - \frac{n-1}{n}\left[b_{0i}V_{ij} + b_{1i} - \sum_{k \neq i} \frac{V_{jk}}{v_k}\right]\right\} \tag{5.C.12}$$

This is Eq. 5.4.3.

DERIVATION OF CONSTRAINED MAXIMUM LIKELIHOOD ESTIMATES (CMLE'S)

The log likelihood L when $\sigma_i^2 = 0$ is found by setting $\sigma_i^2 = 0$ in Eq. 5.B.10:

$$L = C - 0.5n \sum_{k \neq i} \ln \sigma_k^2 - \frac{0.5(n-1) \sum_{k \neq i} V_{ik}}{\sigma_k^2} \qquad (5.D.1)$$

The CMLE's of σ_k^2 for $k \neq i$ are then found by equating to zero the partial derivatives of L with respect to σ_k^2:

$$\frac{\partial L}{\partial \sigma_k^2} = -\frac{0.5n}{\sigma_k^2} + \frac{0.5(n-1)V_{ik}}{\sigma_k^4} = 0 \qquad (5.D.2)$$

from which the CMLE of σ_k^2 is

$$\hat{\sigma}_k^2 = \frac{(n-1)V_{ik}}{n} \qquad (5.D.3)$$

Upon replacing σ_k^2 in the last summation of Eq. 5.D.1 by its CMLE for all $k \neq i$, the sum becomes

$$\frac{\sum_{k \neq i} V_{ik}}{\hat{\sigma}_k^2} = \sum_{k \neq i} \frac{n V_{ik}}{(n-1)V_{ik}}$$

$$= \frac{n(N-1)}{n-1} \qquad (5.D.4)$$

Hence, the σ_i^2 that corresponds to the largest likelihood when one of the σ_k^2's is zero is the one that *minimizes* the first sum in Eq. 5.D.1,

$$\sum_{\substack{k=1 \\ k \neq i}}^{N} \ln \sigma_k^2$$

or, equivalently, that minimizes

$$\prod_{\substack{k=1 \\ k \neq i}}^{N} \sigma_k^2$$

This result is the basis for Eq. 5.5.2.

DERIVATION OF MAXIMUM LIKELIHOOD ESTIMATOR (MLE) OF σ_0^2 AND OF $L(\hat{\omega})$ WHEN PRECISIONS ARE EQUAL

Under H_0: $\sigma_1^2 = \sigma_2^2 = \cdots = \sigma_N^2 = \sigma_0^2$, L of Eq. 5.B.1 reduces to

$$L(\omega) = C - 0.5n \ln Q - \frac{0.5(n-1)(V/Q) \sum\limits_{i=1}^{N-1} \sum\limits_{j>i} V_{ij}}{\sigma_0^4} \qquad (5.E.1)$$

where, from Eqs. 5.4.8 and 5.4.9 and under H_0,

$$V = \sigma_0^{2N} \qquad (5.E.2)$$

$$Q = N\sigma_0^{2(N-1)} \qquad (5.E.3)$$

and

$$\frac{V}{Q} = \frac{\sigma_0^2}{N} \qquad (5.E.4)$$

Thus $L(\omega)$ reduces to

$$L(\omega) = C - 0.5n \ln N - 0.5n(N-1)\ln \sigma_0^2$$

$$- \frac{0.5(n-1) \sum\limits_{i=1}^{N-1} \sum\limits_{j>i} V_{ij}}{N\sigma_0^2} \qquad (5.E.5)$$

To find the MLE of σ_0^2, the partial derivative of $L(\omega)$ is found with respect to

σ_0^2, equated to zero, and solved for σ_0^2.

$$\frac{\partial L(\omega)}{\partial \sigma_0^2} = -\frac{0.5n(N-1)}{\sigma_0^2} + \frac{0.5(n-1)\sum\limits_{i=1}^{N-1}\sum\limits_{j>i} V_{ij}}{N\sigma_0^4} = 0$$

from which

$$\hat{\sigma}_0^2 = \frac{(n-1)\sum\limits_{i=1}^{N-1}\sum\limits_{j>i} V_{ij}}{nN(N-1)} \tag{5.E.6}$$

which is Eq. 5.4.13. At this value of σ_0^2, $L(\omega)$ in Eq. 5.E.5 becomes

$$L(\hat{\omega}) = C - 0.5n \ln N - 0.5n(N-1)\ln \hat{\sigma}_0^2 - 0.5n(N-1)$$

$$= C - 0.5n\left[\ln N + (N-1)\ln \hat{\sigma}_0^2 + (N-1)\right] \tag{5.E.7}$$

which is Eq. 5.4.14.

REFERENCES

1. Grubbs, F. E. "On Estimating Precision of Measuring Instruments and Product Variability." *J. Amer. Stat. Assn.* **43**: 243–264; 1948.
2. Mandel, J. "Models, Transformations of Scale, and Weighting." *J. Quality Technol.* **8**(2): 86–97; 1976.
3. Jaech, J. L. "Estimation of Scale Accuracy and Precision: A Case History." *Nucl. Mater. Manage. J.* **VII**(3): 1978.
4. Mood, A. M. *Introduction to the Theory of Statistics.* New York: McGraw-Hill, 1950.
5. Jaech, J. L. "Large Sample Tests of Instrument Precisions with More Than Two Instruments." *Technometrics.* **18**(2): 127–133; 1976.

Chapter 6

$N \geq 3$ MEASUREMENT METHODS: INFERENCE BASED ON ORIGINAL DATA

6.1 INTRODUCTION

In Chapters 4 and 5, the methods of statistical inference were based on paired differences; that is, the original nN data points x_{ik} $(i = 1, 2, \ldots, N; \; k = 1, 2, \ldots, n)$ were not used in the analysis once y_{ijk} was calculated from Eq. 4.1.2. For the paired difference approach, it made no difference whether the model was random or fixed. (See Section 1.8 for the distinction between the random and fixed models.) Because of this, inferences about μ_k, the true value for item k, could not be made when working with paired differences.

In this chapter, the original x_{ik} values form the basis for the statistical inference. It is necessary to distinguish between the random and fixed models in some instances, and inferences can now be made about σ_μ^2. Further, in working with the original data, inferences can be made about β_i in the model in Eq. 1.8.1. Specifically, the nonconstant bias model, wherein $\beta_i \neq 1$ for some or all i, may now be treated.

Chapter 6 follows much of the outline of the previous two chapters. Grubbs' estimators are given and generalized in Section 6.2. Maximum likelihood estimation is covered in Section 6.3, first for the random and then for the fixed model. Statistical inference forms the basis for Section 6.4, and in Section 6.5, constrained maximum likelihood estimation is presented. Finally, Section 6.6 covers the estimation of relative biases.

Many of the examples in Chapter 6 are those considered in the earlier chapters. The estimates of parameters obtained by the various methods of estimation are compared with one another. Indications as to which estimation methods are the better, in some sense, follow from the discussions on the construction of confidence intervals and on the examples themselves.

Some additional notation is used in this chapter to distinguish among the various estimators. In prior chapters, the tilde (~) was placed over σ^2 to indicate a Grubbs' estimator, and the caret (^) was used for maximum likelihood estimators (MLE's). Further, in Chapters 3 and 4 a dot ($\cdot$) indicated that the estimator was a constrained expected likelihood estimator (CELE). In Chapter 6, a double tilde ($\approx$) is used to denote Grubbs' estimators for $N \geq 3$, when they are derived from the original data rather than from the paired differences, and a double caret ($\hat{\hat{}}$) has the same meaning for MLE's. No distinction is made in notation between estimators derived from the constant bias and nonconstant bias models, but it is always made clear which model is applicable.

6.2 GRUBBS' ESTIMATORS

Grubbs provides estimators only for the constant bias model, that is, for the model

$$x_{ik} = \alpha_i + \mu_k + \varepsilon_{ik} \qquad (6.2.1)$$

For $N = 3$, Grubbs (1) gives two estimators of σ_i^2. One, derived by what he calls Method A_3, is identical to Eq. 4.2.4 in Chapter 4 and is based on paired differences. The other, based on Grubbs' Method B_3 is, for $i = 1$,

$$\tilde{\sigma}_1^2 = s_1^2 - \frac{s_{12} + s_{13} + s_{23}}{3} \qquad (6.2.2)$$

where s_1^2 is calculated from Eq. 1.6.2 for the x_{1k} values, and s_{ij} is the sample covariance between x_{ik} and x_{jk}, calculated by Eq. 2.2.12.

For $N > 3$, Grubbs provides only those estimators based on Method A_N, that is, those that are equivalent to $\tilde{\sigma}_i^2$ given by Eq. 5.2.3. However, Appendix 6.A shows that for $N = 3$, the Method B_3 estimators are similar to the least squares solution of the six moment equations in which s_1^2, s_2^2, s_3^2, s_{12}, s_{13}, and s_{23} are equated to their respective mean squares. Appendix 6.A also shows that for general N, the least squares solution of the $N(N + 1)/2$ moment equations involving the N s_i^2's and the $N(N - 1)/2$ s_{ij}'s is

$$\tilde{\sigma}_i^2 = s_i^2 - \overline{\beta}_i^2 \tilde{\sigma}_\mu^2 \qquad (6.2.3)$$

where

$$\overline{\beta}_i = \frac{\left(\prod_{j \neq i} s_{ij} \right)^{1/N}}{\left(\prod_{\substack{j,k>j \\ \neq i}} s_{jk} \right)^{2/N(N-2)}} \qquad (6.2.4)$$

and

$$\tilde{\sigma}_\mu^2 = \prod_{i,j>i} s_{ij}^{2/N(N-1)} \qquad (6.2.5)$$

The parameter σ_μ^2 is defined in Section 3.2 for both the random and fixed models. The estimator is the same for both models, but the interpretation of σ_μ^2 is model dependent.

For $N = 3$ and $\beta_i = 1$ for all i, Eq. 6.2.3 reduces to the Grubbs' Method B_3 estimator given by Eq. 6.2.2 except for the definition of $\tilde{\sigma}_\mu^2$. Because of this difference, and since Grubbs did not cover the nonconstant bias model, nor did he explicitly provide estimators by Method B_N for $N > 3$ (i.e., estimators based on the original data rather than on the paired differences), we henceforth label all estimators derived by least squares solutions of the moment Eqs. 6.A.1 and 6.A.2 as moment estimators (ME's).

To illustrate, for $N = 5$ and $i = 2$, Eqs. 6.2.4 and 6.2.5 become

$$\tilde{\sigma}_\mu^2 = \left(s_{12} s_{13} \cdots s_{45} \right)^{0.1}$$

$$\overline{\beta}_2 = \frac{\left(s_{12} s_{23} s_{24} s_{25} \right)^{1/5}}{\left(s_{13} s_{14} s_{15} s_{34} s_{35} s_{45} \right)^{2/15}}$$

6.3 MAXIMUM LIKELIHOOD ESTIMATORS (MLE'S)

The MLE's of σ_i^2 and of σ_μ^2 for the random model are presented in Section 6.3.1. Maximum likelihood estimation for the fixed model is discussed in Section 6.3.2.

6.3.1 Maximum Likelihood Estimators (MLE's) for the Random Model

The MLE's for the random model are derived in Appendix 6.B for the most general nonconstant bias case, in which $\beta_i \neq 1$ for all i. For the constant bias model, the parameter estimates are found by setting $\beta_i = 1$ for all i in the estimating equations to follow. If desired, intermediate cases in which $\beta_i = 1$ for some but not all i can also be handled by the general formulation given here.

The estimating equations for finding the MLE's of σ_i^2 and of σ_μ^2 require that β_i be known for all i. Since β_i is usually not known in the general nonconstant-bias model case, the estimator $\overline{\beta}_i$ in Eq. 6.2.4 is used. The principle of maximum likelihood estimation applies only to the parameters σ_μ^2 and σ_i^2 for $i = 1, 2, \ldots, N$ in the presentation to follow.

It is shown in Appendix 6.B that the MLE of σ_i^2 is

$$\hat{\sigma}_i^2 = \frac{(n-1)(b_{1i} b_{2i} - b_{0i} b_{3i})}{n b_{1i}^2} - \frac{b_{0i}}{b_{1i}} \tag{6.3.1}$$

where, after defining

$$a_i = \bar{\beta}_i \hat{\sigma}_\mu \qquad (6.3.2)$$

the b_{ki}'s are

$$b_{0i} = a_i^2 \qquad (6.3.3)$$

$$b_{1i} = \sum_{j \neq i} \frac{a_j^2}{\sigma_j^2} + 1 \qquad (6.3.4)$$

$$b_{2i} = b_{1i}s_i^2 + a_i^2 \sum_{j \neq i} \frac{s_j^2}{\sigma_j^2} - 2a_i \sum_{j \neq i} \frac{a_j s_{ij}}{\sigma_j^2} \qquad (6.3.5)$$

$$b_{3i} = b_{1i} \sum_{j \neq i} \frac{s_j^2}{\sigma_j^2} - \sum_{j \neq i} \frac{a_j^2 s_j^2}{\sigma_j^4} - 2 \sum_{j \neq i} \sum_{\substack{k > j \\ \neq i}} \frac{a_j a_k s_{jk}}{\sigma_j^2 \sigma_k^2} \qquad (6.3.6)$$

and $\hat{\sigma}_\mu^2$ is the MLE of σ_μ^2 given by

$$\hat{\sigma}_\mu^2 = \frac{(n-1)(d_2 d_3 - d_4)}{n d_2^2} - \frac{1}{d_2} \qquad (6.3.7)$$

where

$$d_2 = \sum_{i=1}^{N} \frac{\bar{\beta}_i^2}{\sigma_i^2} \qquad (6.3.8)$$

$$d_3 = \sum_{i=1}^{N} \frac{s_i^2}{\sigma_i^2} \qquad (6.3.9)$$

$$d_4 = \sum_{i}^{N} \sum_{j \neq i}^{N} \frac{\bar{\beta}_i^2 s_j^2}{\sigma_i^2 \sigma_j^2} - 2 \sum_{i}^{N-1} \sum_{j > i}^{N} \frac{\bar{\beta}_i \bar{\beta}_j s_{ij}}{\sigma_i^2 \sigma_j^2} \qquad (6.3.10)$$

As in Chapter 5, an iterative estimation procedure is used. Initial values are assigned to $\sigma_1^2, \sigma_2^2, \ldots, \sigma_N^2$, and the first estimate of σ_μ^2 is given by Eq. 6.3.7. This is then used with the initial values of $\sigma_2^2, \ldots, \sigma_N^2$ to calculate the initial estimate of σ_1^2. The procedure continues until the results of a given iteration agree, within specified criteria, with those of the previous iteration. The estimation procedure is easily programmed, and a Fortran program computer listing is given in Chapter 9.

6.3.2 Maximum Likelihood Estimators (MLE's) for the Fixed Model

For a fixed model, the aim is to obtain estimates of μ_k for each k rather than to obtain estimates of σ_μ^2. Unfortunately, this is impossible, since the MLE is undefined for the fixed model. This fact was reported by Anderson and Rubin (2), who suggest that the parameter estimators for the random model be used as well for the fixed model. Further consideration of the fixed model is contained in Section 6.5.2.

A number of examples considered in Chapters 4 and 5 are reconsidered here to compare the various estimators.

EXAMPLE 6.A

Reconsider Example 4.A dealing with measuring 43 pellet densities by Methods 1, 3, and 6. The six moments are found to be

$$s_1^2 = 0.044954 \qquad s_{13} = 0.040267$$

$$s_3^2 = 0.065953 \qquad s_{16} = 0.034938$$

$$s_6^2 = 0.069424 \qquad s_{36} = 0.034433$$

For the constant bias model, $\beta_1 = \beta_3 = \beta_6 = 1$. From Eq. 6.2.5,

$$\tilde{\sigma}_\mu^2 = 0.036454$$

and from Eq. 6.2.3,

$$\tilde{\sigma}_1^2 = 0.008500 \qquad \tilde{\sigma}_3^2 = 0.029499 \qquad \tilde{\sigma}_6^2 = 0.032970$$

The MLE's for the constant bias model are found by iteration on Eqs. 6.3.1 and 6.3.7. The results of selected iterations are listed in Table 6.1.

Table 6.1 MLE's for Example 6.A in Each Iteration

Iteration	σ_1^2	σ_3^2	σ_6^2	σ_μ^2
Input	0.0042	0.0262	0.0403	—
1	0.004294	0.025404	0.038397	0.048479
2	0.004639	0.025178	0.038158	0.048702
⋮	⋮	⋮	⋮	⋮
5	0.004704	0.025135	0.038123	0.049087
⋮	⋮	⋮	⋮	⋮
7 and 8	0.004704	0.025135	0.038123	0.049088

Table 6.2   Parameter Estimates for Example 6.A, NCB Model

Iteration	σ_1^2	σ_3^2	σ_6^2	σ_μ^2
Input	0.0042	0.0262	0.0403	—
1	0.003035	0.026379	0.039117	0.044349
2	0.003164	0.026271	0.039041	0.043164
⋮	⋮	⋮	⋮	⋮
5	0.003195	0.026246	0.039024	0.043354
6	0.003195	0.026246	0.039024	0.043355

For the nonconstant bias model, from Eq. 6.2.4,

$$\bar{\beta}_1 = \frac{[(0.040267)(0.034938)]^{1/3}}{(0.034433)^{2/3}} = 1.0587$$

$$\bar{\beta}_3 = 1.0434 \qquad \bar{\beta}_6 = 0.9053$$

From Eq. 6.2.3,

$$\tilde{\sigma}_1^2 = 0.044954 - (1.0587)^2(0.036454) = 0.004095$$

$$\tilde{\sigma}_3^2 = 0.026266 \qquad \tilde{\sigma}_6^2 = 0.039547$$

The MLE's are again found by iteration on Eqs. 6.3.1 and 6.3.7, this time using $\bar{\beta}_i$ in place of unity. Table 6.2 shows the results.

In the succeeding tables, methods are identified as shown in this list and in Table 6.3:

ME = Moment estimators (Grubbs or modified Grubbs)
MLE = Maximum likelihood estimators

Table 6.3   Estimation Methods Identification

Identification	Equations
ME-PD	5.2.3
MLE-PD	5.3.1
ME-OD-CB	6.2.3, 6.2.5
MLE-OD-CB	6.3.1, 6.3.7
ME-OD-NCB	6.2.3–6.2.5
MLE-OD-NCB	6.3.1, 6.3.7, 6.2.4

Table 6.4 Parameter Estimates for Example 6.A ($\times 10^{-6}$)

| Parameter | PD | | OD | | | |
| | | | CB | | NCB | |
	ME	MLE	ME	MLE	ME	MLE
σ_1^2	4,183	4,086	8,500	4,704	4,095	3,195
σ_3^2	26,192	25,583	29,499	25,135	26,266	26,246
σ_6^2	40,320	39,382	32,970	38,123	39,547	39,024
σ_μ^2	—	—	36,454	49,088	36,454	43,355

$$\bar{\beta}_1 = 1.058683 \qquad \bar{\beta}_3 = 1.043380 \qquad \bar{\beta}_6 = 0.905298$$

PD = Paired differences
OD = Original data
CB = Constant bias ($\beta_i = 1$)
NCB = Nonconstant bias ($\beta_i \neq 1$)

Table 6.4 summarizes all the estimates for this set of data.

For the next four examples, $N = 3$. Only the summary tables analogous to Table 6.4 are given. ∎

EXAMPLE 6.B

Reconsider Example 4.B dealing with the image extinction method for measuring Cerenkov glow intensity of 36 items, with measurements made by 3 observers. The moments are

$$s_1^2 = 0.104349 \qquad s_{12} = 0.037002$$

$$s_2^2 = 0.077309 \qquad s_{13} = 0.028931$$

$$s_3^2 = 0.055511 \qquad s_{23} = 0.011352$$

Table 6.5 summarizes the estimates.

With the slope estimates differing so greatly from unity in this example, it would appear that the estimates based on the nonconstant bias model are the appropriate ones. In Section 6.4.2, methods for testing whether $\beta_i = 1$ for all i are provided. ∎

EXAMPLE 6.C

Reconsider Example 4.C dealing with Grubbs' original fuse burning data (1) in which 3 observers measured elapsed times for 30 fuses (one incomplete set of

Table 6.5   Parameter Estimates for Example 6.B ($\times 10^{-10}$)

| | PD | | OD | | | |
| | | | CB | | NCB | |
Parameter	ME	MLE	ME	MLE	ME	MLE
σ_1^2	497.69	483.87	813.58	578.05	100.53	302.67
σ_2^2	578.86	562.78	543.18	542.37	627.89	620.14
σ_3^2	522.30	507.79	325.20	381.24	466.36	459.30
σ_μ^2	—	—	229.91	735.30	229.91	257.72
	$\bar{\beta}_1 = 2.0252$		$\bar{\beta}_2 = 0.7947$		$\bar{\beta}_3 = 0.6213$	

readings was discarded). The six moments are

$$s_1^2 = 0.046754 \qquad s_{12} = 0.045582$$

$$s_2^2 = 0.045112 \qquad s_{13} = 0.045253$$

$$s_3^2 = 0.044640 \qquad s_{23} = 0.044721$$

Table 6.6 summarizes the estimates.   ■

EXAMPLE 6.D

Reconsider Example 4.D dealing with the measurement of stress at 600% elongation of seven rubber samples by Laboratories 6, 8, and 12. For the

Table 6.6   Parameter Estimates for Example 6.C ($\times 10^{-6}$)

| | PD | | OD | | | |
| | | | CB | | NCB | |
Parameter	ME	MLE	ME	MLE	ME	MLE
σ_1^2	641	619	1,570	619	625	609
σ_2^2	63	61	− 72	62	63	64
σ_3^2	249	240	− 544	237	239	234
σ_μ^2	—	—	45,184	43,586	45,184	43,764
	$\bar{\beta}_1 = 1.0104$		$\bar{\beta}_2 = 0.9985$		$\bar{\beta}_3 = 0.9913$	

Table 6.7 Parameter Estimates for Example 6.D $(\times 10^{-6})$

| Parameter | PD | | OD | | | |
| | ME | MLE | CB | | NCB | |
			ME	MLE	ME	MLE
σ_6^2	1,019	873	−3,676	914	1,088	943
σ_8^2	724	621	−8,818	570	614	533
σ_{12}^2	4,430	3,797	18,809	3,838	4,042	3,440
σ_μ^2	—	—	281,428	237,201	281,428	242,137
	$\bar{\beta}_6 = 0.9915$		$\bar{\beta}_8 = 0.9831$		$\bar{\beta}_{12} = 1.0259$	

logarithmically transformed data, the six moments are

$$s_6^2 = 0.277752 \qquad s_{68} = 0.274310$$

$$s_8^2 = 0.272610 \qquad s_{6,12} = 0.286270$$

$$s_{12}^2 = 0.300237 \qquad s_{8,12} = 0.283847$$

Table 6.7 summarizes the estimates. ∎

EXAMPLE 6.E

Reconsider Example 4.N dealing with the measurement of automobile fuel consumption. The six moments are:

$$s_1^2 = 21.598460 \qquad s_{12} = 23.546331$$

$$s_2^2 = 25.779234 \qquad s_{13} = 23.843730$$

$$s_3^2 = 26.479947 \qquad s_{23} = 26.109789$$

Table 6.8 summarizes the estimates. ∎

A number of additional examples for which $N > 3$ are now given. Rather detailed results are presented for Example 6.F to illustrate the calculations. For succeeding examples, only the summary tables are given, except in those cases where the data are not available from previous chapters. In such cases, the data are also listed.

Table 6.8   Parameter Estimates for Example 6.E

	PD		OD			
			CB		NCB	
Parameter	ME	MLE	ME	MLE	ME	MLE
σ_1^2	0.3182	0.2983	−2.8753	0.2972	0.0975	0.0898
σ_2^2	−0.0332	−0.0311	1.3055	−0.0273	−0.0038	−0.0046
σ_3^2	0.0728	0.0682	2.0062	0.0635	0.0397	0.0379
σ_μ^2	—	—	24.4737	24.3306	24.4737	22.9388

$$\bar{\beta}_1 = 0.9373 \qquad \bar{\beta}_2 = 1.0264 \qquad \bar{\beta}_3 = 1.0394$$

EXAMPLE 6.F

Reconsider Example 5.A dealing with the measurements of 43 pellet densities by each of 6 methods. The 21 moments are calculated to be

$$s_1^2 = 0.044954 \qquad s_{12} = 0.035313 \qquad s_{26} = 0.041667$$
$$s_2^2 = 0.041083 \qquad s_{13} = 0.040267 \qquad s_{34} = 0.051683$$
$$s_3^2 = 0.065953 \qquad s_{14} = 0.050293 \qquad s_{35} = 0.048413$$
$$s_4^2 = 0.128748 \qquad s_{15} = 0.038996 \qquad s_{36} = 0.034433$$
$$s_5^2 = 0.149426 \qquad s_{16} = 0.034938 \qquad s_{45} = 0.066600$$
$$s_6^2 = 0.069424 \qquad s_{23} = 0.035661 \qquad s_{46} = 0.038081$$
$$\qquad\qquad\qquad s_{24} = 0.036551 \qquad s_{56} = 0.023252$$
$$\qquad\qquad\qquad s_{25} = 0.024125$$

First assume a constant bias model. From Eq. 6.2.5,

$$\tilde{\sigma}_\mu^2 = [(0.035313)(0.040267)\ldots(0.023252)]^{1/15} = 38{,}678 \times 10^{-6}$$

From Eq. 6.2.3, for $\beta_i = 1$ for all i,

$$\tilde{\sigma}_1^2 = (44{,}954 - 38{,}678) \times 10^{-6} = 6276 \times 10^{-6}$$

$$\tilde{\sigma}_2^2 = 2405 \times 10^{-6} \qquad \tilde{\sigma}_3^2 = 27{,}275 \times 10^{-6}$$

$$\tilde{\sigma}_4^2 = 90{,}070 \times 10^{-6} \qquad \tilde{\sigma}_5^2 = 110{,}748 \times 10^{-6} \qquad \tilde{\sigma}_6^2 = 30{,}746 \times 10^{-6}$$

The MLE's are found by iteration on Eqs. 6.3.1 and 6.3.7 with $\beta_i = 1$ for all i. Table 6.9 gives the parameter estimates for selected iterations.

Table 6.9 MLE's for Example 6.F in Each Iteration ($\times 10^{-6}$)

Iteration	σ_1^2	σ_2^2	σ_3^2	σ_4^2	σ_5^2	σ_6^2	σ_μ^2
Input	5,000	14,000	21,000	67,000	109,000	43,000	—
1	5,037	8,855	24,157	70,950	113,081	34,560	53,243
2	6,130	8,126	24,738	72,968	114,054	32,753	51,740
⋮	⋮	⋮	⋮	⋮	⋮	⋮	⋮
5	6,551	7,826	24,839	74,342	114,555	31,222	52,175
⋮	⋮	⋮	⋮	⋮	⋮	⋮	⋮
7	6,563	7,818	24,849	74,383	114,570	31,186	52,178

For the nonconstant bias model, from Eq. 6.2.4,

$$\bar{\beta}_1 = \frac{[(0.035313)(0.040267)\ldots(0.034938)]^{1/6}}{[(0.035661)(0.036551)\ldots(0.023252)]^{1/12}} = 1.0300$$

$$\bar{\beta}_2 = 0.8550 \qquad \bar{\beta}_3 = 1.0934$$

$$\bar{\beta}_4 = 1.2917 \qquad \bar{\beta}_5 = 0.9501 \qquad \bar{\beta}_6 = 0.8462$$

From Eq. 6.2.3,

$$\tilde{\sigma}_1^2 = \left[44{,}954 - (1.0300)^2 38{,}678\right] \times 10^{-6} = 3921 \times 10^{-6}$$

$$\tilde{\sigma}_2^2 = 12{,}808 \times 10^{-6} \qquad \tilde{\sigma}_3^2 = 19{,}713 \times 10^{-6}$$

$$\tilde{\sigma}_4^2 = 64{,}214 \times 10^{-6} \qquad \tilde{\sigma}_5^2 = 114{,}512 \times 10^{-6} \qquad \tilde{\sigma}_6^2 = 41{,}728 \times 10^{-6}$$

The MLE's are found by iteration on Eqs. 6.3.1 and 6.3.7 with $\beta_i = \bar{\beta}_i$ given above. Table 6.10 gives the parameter estimates for selected iterations. Table 6.11 summarizes the estimates. ∎

EXAMPLE 6.G

Reconsider Example 5.B dealing with the measurement of stress at 600% elongation of 7 rubber samples by Laboratories 3, 5, 6, 8, and 12. (See also Example 6.D.) For the logarithmically transformed data, the nine moments not

Table 6.10   MLE's for Example 6.F, NCB Model ($\times 10^{-6}$)

Iteration	σ_1^2	σ_2^2	σ_3^2	σ_4^2	σ_5^2	σ_6^2	σ_μ^2
Input	5,000	14,000	21,000	67,000	109,000	43,000	—
1	4,253	10,089	23,692	66,589	110,779	36,162	52,164
2	4,681	9,906	24,427	68,141	110,939	35,572	51,554
⋮	⋮	⋮	⋮	⋮	⋮	⋮	⋮
5	4,759	9,869	24,170	68,260	110,949	35,195	52,253
6	4,759	9,869	24,169	68,261	110,949	35,194	52,256

Table 6.11   Parameter Estimates for Example 6.F ($\times 10^{-6}$)

	PD		OD			
			CB		NCB	
Parameter	ME	MLE	ME	MLE	ME	MLE
σ_1^2	5,077	6,151	6,276	6,563	3,921	4,759
σ_2^2	14,457	8,628	2,405	7,818	12,808	9,869
σ_3^2	20,752	24,503	27,275	24,849	19,713	24,169
σ_4^2	67,171	72,838	90,070	74,383	64,214	68,261
σ_5^2	108,759	113,986	110,748	114,570	114,512	110,949
σ_6^2	43,269	31,918	30,746	31,186	41,728	35,194
σ_μ^2	—	—	38,678	52,178	38,678	52,256

$$\bar{\beta}_1 = 1.0300 \qquad \bar{\beta}_4 = 1.2917$$
$$\bar{\beta}_2 = 0.8550 \qquad \bar{\beta}_5 = 0.9501$$
$$\bar{\beta}_3 = 1.0934 \qquad \bar{\beta}_6 = 0.8462$$

previously given in Example 6.D are as follows:

$$s_3^2 = 0.298384 \qquad s_{35} = 0.310609 \qquad s_{56} = 0.299416$$
$$s_5^2 = 0.325591 \qquad s_{36} = 0.285522 \qquad s_{58} = 0.294973$$
$$\qquad\qquad\qquad\quad s_{38} = 0.281955 \qquad s_{5,12} = 0.311068$$
$$\qquad\qquad\qquad\quad s_{3,12} = 0.298810$$

Table 6.12 summarizes the estimates.   ∎

EXAMPLE 6.H

Reconsider Example 5.C dealing with the measurements of uranium concentration on each of eight samples by each of four laboratories. The 10 moments are

Table 6.12   Parameter Estimates for Example 6.G $(\times 10^{-6})$

| | PD | | OD | | | |
| | | | CB | | NCB | |
Parameter	ME	MLE	ME	MLE	ME	MLE
σ_3^2	1,583	477	5,706	479	1,535	566
σ_5^2	2,678	2,146	32,913	2,165	1,252	1,451
σ_6^2	1,870	3,442	$-14,926$	3,420	1,348	2,645
σ_8^2	3,683	4,386	$-20,068$	4,359	2,766	3,482
σ_{12}^2	1,370	531	7,559	533	1,203	667
σ_μ^2	—	—	292,678	256,468	292,678	251,948

$$\bar{\beta}_3 = 1.0071 \qquad \bar{\beta}_8 = 0.9602$$
$$\bar{\beta}_5 = 1.0527 \qquad \bar{\beta}_{12} = 1.0108$$
$$\bar{\beta}_6 = 0.9718$$

Table 6.13   Parameter Estimates for Example 6.H $(\% \ U)^2$

| | PD | | OD | | | |
| | | | CB | | NCB | |
Parameter	ME	MLE	ME	MLE	ME	MLE
σ_1^2	5.3838	4.6655	7.7910	4.5941	5.3827	5.1668
σ_2^2	0.0606	1.0505	-0.8145	0.9624	0.0531	1.5946
σ_3^2	0.5142	0.4486	10.2935	0.5442	0.0490	-0.1200
σ_4^2	2.5224	1.2059	-8.7890	1.1862	2.0145	0.7475
σ_μ^2	—	—	117.4517	105.5165	117.4517	102.5657

$$\bar{\beta}_1 = 1.0102 \qquad \bar{\beta}_3 = 1.0427$$
$$\bar{\beta}_2 = 0.9963 \qquad \bar{\beta}_4 = 0.9529$$

Table 6.14   Parameter Estimates for Example 6.I (grams^2)

| | PD | | OD | | | |
| | | | CB | | NCB | |
Parameter	ME	MLE	ME	MLE	ME	MLE
σ_1^2	300.92	210.57	$-121,534$	209.73	2,071	1.99
σ_2^2	202.92	140.63	52,321	141.38	$-1,509$	73.25
σ_3^2	1,014.58	1,106.31	95,935	1,106.64	-998	1,104.14
σ_4^2	809.92	644.43	78,063	644.88	2,686	546.09
σ_5^2	633.92	757.34	$-101,521$	756.47	608	857.38
σ_μ^2	—	—	26,901,598	25,100,037	26,901,598	25,108,254

$$\bar{\beta}_1 = 0.9977 \qquad \bar{\beta}_3 = 1.0018 \qquad \bar{\beta}_5 = 0.9981$$
$$\bar{\beta}_2 = 1.0010 \qquad \bar{\beta}_4 = 1.0014$$

as follows, in $(\% \ U)^2$.

$$s_1^2 = 125.242670 \qquad s_{12} = 118.929234 \qquad s_{23} = 121.227945$$
$$s_2^2 = 116.637255 \qquad s_{13} = 123.509395 \qquad s_{24} = 111.323004$$
$$s_3^2 = 127.745255 \qquad s_{14} = 112.323689 \qquad s_{34} = 117.397196$$
$$s_4^2 = 108.662793$$

Table 6.13 summarizes the estimates.   ■

EXAMPLE 6.I

Reconsider Example 5.D dealing with the measurements of the mass of 15 items on each of 5 scales. The 15 moments in kg^2 are as follows:

$$s_1^2 = 26.780064 \qquad s_{12} = 26.866810 \qquad s_{24} = 26.966407$$
$$s_2^2 = 26.953919 \qquad s_{13} = 26.887988 \qquad s_{25} = 26.876450$$
$$s_3^2 = 26.997533 \qquad s_{14} = 26.879386 \qquad s_{34} = 26.987629$$
$$s_4^2 = 26.979661 \qquad s_{15} = 26.789605 \qquad s_{35} = 26.898255$$
$$s_5^2 = 26.800077 \qquad s_{23} = 26.975051 \qquad s_{45} = 26.889000$$

Table 6.14 summarizes the results. All variances are in $grams^2$.   ■

EXAMPLE 6.J

Reconsider Example 5.E dealing with the measurements of the mass of 39 items on each of 14 scales. The 105 moments are listed.

i	s_i^2
1	58.565022
2	58.549600
3	58.585994
4	58.533720
5	58.526450
6	58.572237
7	58.537800
8	58.499550
9	58.512360
10	58.517805
11	58.519387
12	58.445082
13	58.534626
14	58.546202

i,j	s_{ij}	i,j	s_{ij}	i,j	s_{ij}	i,j	s_{ij}
1,2	58.557296	2,12	58.497272	4,14	58.539919	7,13	58.536160
1,3	58.575460	2,13	58.542078	5,6	58.549322	7,14	58.541954
1,4	58.549335	2,14	58.547870	5,7	58.532091	8,9	58.505914
1,5	58.545703	3,4	58.559838	5,8	58.512978	8,10	58.508667
1,6	58.568598	3,5	58.556185	5,9	58.519364	8,11	58.509431
1,7	58.551371	3,6	58.579066	5,10	58.522099	8,12	58.472271
1,8	58.532256	3,7	58.561873	5,11	58.522865	8,13	58.517045
1,9	58.538643	3,8	58.542737	5,12	58.485688	8,14	58.522835
1,10	58.541373	3,9	58.549133	5,13	58.530483	9,10	58.515019
1,11	58.542180	3,10	58.551864	5,14	58.536292	9,11	58.515824
1,12	58.504977	3,11	58.552625	6,7	58.554962	9,12	58.478668
1,13	58.549795	3,12	58.515440	6,8	58.535852	9,13	58.523435
1,14	58.555591	3,13	58.560248	6,9	58.542242	9,14	58.529234
2,3	58.567742	3,14	58.566049	6,10	58.544977	10,11	58.518544
2,4	58.541622	4,5	58.530065	6,11	58.545742	10,12	58.481378
2,5	58.537986	4,6	58.552942	6,12	58.508553	10,13	58.526161
2,6	58.560876	4,7	58.535738	6,13	58.553367	10,14	58.531957
2,7	58.543653	4,8	58.516616	6,14	58.559180	11,12	58.482174
2,8	58.524543	4,9	58.523001	7,8	58.518654	11,13	58.526981
2,9	58.530934	4,10	58.525746	7,9	58.525033	11,14	58.532761
2,10	58.533656	4,11	58.526507	7,10	58.527777	12,13	58.489777
2,11	58.534366	4,12	58.489334	7,11	58.528545	12,14	58.495560
		4,13	58.534120	7,12	58.491380	13,14	58.540391

Table 6.15 gives the ME's for the paired differences and the MLE's for all cases. Note that the MLE's based on paired differences and those based on the original data agree to within 0.01 for the constant bias model. ∎

EXAMPLE 6.K

Example 6.G is extended to include all 13 laboratories that measured the stress at 600% elongation of the 7 rubber samples. The data for Laboratories 3, 5, 6, 8, and 12 were given in Example 5.B. The logarithmically transformed data for the remaining eight laboratories are given in Table 6.16.

The moments are not listed, nor are the ME's. The MLE's are given in Table 6.17 for the various models. The parenthetical values are taken from Table 6.12, which listed data from only the five laboratories.

With reference to Example 6.G, note that Laboratories 3, 5, 6, 8, and 12 have β_i values that are not too different, and hence, the MLE's based on the constant bias and nonconstant bias models are in reasonably close agreement in Table 6.12. On the other hand, for Laboratories 2, 4, and 7, $\bar{\beta}_i$ differs considerably from 1.00, and as a consequence, the estimates derived from the nonconstant bias model are quite different from those based on the constant bias model. ∎

Table 6.15   Parameter Estimates for Example 6.J (grams^2)

Parameter	PD		OD	
			CB	NCB
	ME	MLE	MLE	MLE
σ_1^2	27.62	30.51	30.51	26.89
σ_2^2	35.53	37.89	37.89	37.59
σ_3^2	52.45	46.22	46.23	32.85
σ_4^2	24.70	21.89	21.89	21.45
σ_5^2	27.28	28.14	28.14	28.27
σ_6^2	57.95	52.01	52.01	45.89
σ_7^2	36.03	33.24	33.24	31.78
σ_8^2	18.95	17.89	17.88	11.40
σ_9^2	54.78	57.48	57.48	56.23
σ_{10}^2	37.03	30.18	30.18	27.50
σ_{11}^2	47.45	50.71	50.71	51.33
σ_{12}^2	104.62	100.58	100.58	67.29
σ_{13}^2	53.62	56.52	56.52	58.11
σ_{14}^2	38.62	41.64	41.64	42.93
σ_μ^2	—	—	57,031,166	57,031,010

$$\bar{\beta}_1 = 1.0003 \qquad \bar{\beta}_6 = 1.0003 \qquad \bar{\beta}_{11} = 0.9999$$
$$\bar{\beta}_2 = 1.0002 \qquad \bar{\beta}_7 = 1.0001 \qquad \bar{\beta}_{12} = 0.9999$$
$$\bar{\beta}_3 = 1.0005 \qquad \bar{\beta}_8 = 0.9997 \qquad \bar{\beta}_{13} = 0.9993$$
$$\bar{\beta}_4 = 1.0000 \qquad \bar{\beta}_9 = 0.9998 \qquad \bar{\beta}_{14} = 1.0000$$
$$\bar{\beta}_5 = 1.0000 \qquad \bar{\beta}_{10} = 0.9999$$

Table 6.16   Logarithm of Stress of Rubber Samples

Sample	Lab 1	Lab 2	Lab 4	Lab 7	Lab 9	Lab 10	Lab 11	Lab 13
1	4.255	4.039	3.692	3.729	3.828	4.308	4.123	3.907
2	4.903	4.861	4.097	4.285	4.252	4.903	4.718	4.461
3	3.408	3.083	3.196	3.275	3.239	3.706	3.320	3.198
4	3.985	3.990	3.631	3.669	3.664	4.227	3.905	3.778
5	3.330	3.277	3.055	3.127	3.188	3.545	3.015	3.108
6	3.373	3.197	3.102	3.158	3.157	3.624	3.146	3.156
7	3.661	3.582	3.400	3.400	3.442	3.920	3.588	3.466

EXAMPLE 6.L

In this example, 15 line segments varying in length from $\frac{7}{8}$ in. to $4\frac{7}{16}$ in. were presented in random order to each of 6 observers, who were asked to estimate the length of each line segment to the nearest $\frac{1}{16}$ in. In Table 6.18, Methods 1 to 6 correspond to the six observers, while Method 7 corresponds to the line segment lengths measured to the nearest $\frac{1}{16}$ in.

Table 6.17 Parameter Estimates for Example 6.K ($\times 10^{-6}$) (MLE's)

Parameter	PD		OD-CB		OD-NCB	
σ_1^2	4,796		4,964		2,935	
σ_2^2	15,103		15,487		9,218	
σ_3^2	909	(477)	998	(479)	1,063	(566)
σ_4^2	23,122		22,470		572	
σ_5^2	3,007	(2,146)	3,116	(2,165)	2,702	(1,451)
σ_6^2	3,388	(3,442)	3,214	(3,420)	3,729	(2,645)
σ_7^2	16,198		15,582		1,538	
σ_8^2	3,488	(4,386)	3,300	(4,359)	3,254	(3,482)
σ_9^2	16,549		16,018		540	
σ_{10}^2	3,758		3,489		754	
σ_{11}^2	6,569		6,864		2,966	
σ_{12}^2	343	(531)	433	(533)	99	(667)
σ_{13}^2	2,006		1,864		229	
σ_μ^2	—		248,527	(256,468)	218,408	

$$\bar{\beta}_1 = 1.1379 \quad \bar{\beta}_6 = 1.0335 \quad \bar{\beta}_{11} = 1.1843$$
$$\bar{\beta}_2 = 1.2200 \quad \bar{\beta}_7 = 0.8066 \quad \bar{\beta}_{12} = 1.0800$$
$$\bar{\beta}_3 = 1.0742 \quad \bar{\beta}_8 = 1.0247 \quad \bar{\beta}_{13} = 0.9799$$
$$\bar{\beta}_4 = 0.7418 \quad \bar{\beta}_9 = 0.7934$$
$$\bar{\beta}_5 = 1.1204 \quad \bar{\beta}_{10} = 0.9500$$

Table 6.18 Line Segment Lengths, Example 6.L ($\frac{1}{16}$ in.)

Method 1	2	3	4	5	6	7
46	45	40	37	48	32	43
18	18	16	16	16	16	19
35	45	36	38	44	32	41
61	67	56	56	64	64	64
19	22	20	20	24	24	25
17	15	15	14	15	16	17
36	45	36	32	40	40	42
69	80	68	58	64	64	71
40	45	42	34	40	40	43
64	80	64	58	64	64	69
15	13	12	11	12	16	14
28	29	29	24	24	32	27
15	13	10	12	12	16	14
38	45	32	36	40	40	38
60	66	52	52	56	56	61

Table 6.19   Parameter Estimates for Example 6.L $(\frac{1}{16}$ in.$)^2$,
              NCB Model—MLE's

	Methods 1 to 6	Methods 1 to 7
σ_1^2	5.69	6.04
σ_2^2	3.38	3.41
σ_3^2	6.36	5.37
σ_4^2	2.39	2.97
σ_5^2	8.49	8.39
σ_6^2	14.17	13.56
σ_7^2	—	0.73
$\bar{\beta}_1$	0.9970	0.9897
$\bar{\beta}_2$	1.2321	1.2233
$\bar{\beta}_3$	0.9788	0.9723
$\bar{\beta}_4$	0.8895	0.8829
$\bar{\beta}_5$	1.0058	0.9995
$\bar{\beta}_6$	0.9297	0.9238
$\bar{\beta}_7$	—	1.0420
σ_μ^2	334.46	340.33

It is quite apparent from Table 6.18 that the nonconstant bias model applies, and the MLE's for this model are the only ones given in Table 6.19. In obtaining these MLE's, Method 7 is first excluded from the data and then included as part of the data set.

Note the small estimate for σ_7^2 in Table 6.19. This is to be expected, since Method 7 corresponds to actual measured values.   ∎

The examples illustrate that there may be considerable differences among the parameter estimates depending upon the assumed model and the estimation method (ME or MLE). It is quite apparent that unless σ_μ^2 is very small, the ME's based on the original data for the constant bias model will produce poor estimates, and these estimators are no longer considered in this development. Further comparisons of estimators are made in the next section.

6.4   INFERENCE

Statistical inference problems include both the construction of interval estimates and the testing of hypotheses. Section 6.4.1 discusses interval estimation, while Section 6.4.2 considers problems in hypothesis testing.

6.4.1 Confidence Intervals on σ_i^2

Moment Estimation. Although the moment estimation method is applicable to either a random or fixed model or to gradations of models between these two extremes, the method given here for constructing confidence intervals assumes a random model, that is, μ_k is randomly selected from a normal population with mean μ and variance σ_μ^2. For this case, Jaech (3) gives the expression for the variance of $\tilde{\sigma}_i^2$ for the nonconstant bias model:

$$\operatorname{var}\tilde{\sigma}_i^2 = \frac{2}{n-1}\left(\sigma_i^4 + 2\beta_i^4\left\{\left[(N-3)/(N-1)(N-2)^2\right]\left(\sigma_i^2/\beta_i^2\right)\sum_{\substack{j=1\\ \neq i}}^{N}\frac{\sigma_j^2}{\beta_j^2}\right.\right.$$
$$\left.\left. +\left[1/(N-1)^2(N-2)^2\right]\sum_{j=1}^{N-1}\sum_{l>j}\frac{\sigma_j^2\sigma_l^2}{\beta_j^2\beta_l^2}\right\}\right) \quad (6.4.1)$$

Note that var $\tilde{\sigma}_i^2$ is independent of σ_μ^2 for the nonconstant bias model, unlike the constant bias model (1). As in previous applications, the unknown parameter values are replaced by their estimates in evaluating var $\tilde{\sigma}_i^2$. Once var $\tilde{\sigma}_i^2$ has been calculated, Eqs. 3.4.2 and 3.4.3 may be applied to find the approximate number of degrees of freedom associated with $\tilde{\sigma}_i^2$ and to construct the appropriate confidence interval. First, however, the construction of intervals based on the MLE's is described.

Confidence Intervals Based on MLE's, Random Model. The general approach is the same as that described in Chapter 5 for the constant model. The H matrix is a square symmetric matrix of dimension $(N + 1)$ with the following elements:

Position (i, i): $\qquad -E\left[\dfrac{\partial^2 L}{\partial\left(\sigma_i^2\right)^2}\right];\qquad i = 1, 2, \ldots, N$

Position (i, j): $\qquad -E\left[\dfrac{\partial^2 L}{\partial\sigma_i^2\,\partial\sigma_j^2}\right];\qquad \begin{array}{l} i = 1, 2, \ldots, N-1 \\ j > i \end{array}$

Position $(N+1, N+1)$: $\quad -E\left[\dfrac{\partial^2 L}{\partial\left(\sigma_\mu^2\right)^2}\right]$

Position $(i, N+1)$: $\qquad -E\left(\dfrac{\partial^2 L}{\partial\sigma_\mu^2\,\partial\sigma_i^2}\right);\qquad i = 1, 2, \ldots, N$

$$(6.4.2)$$

where L is given by Eq. 6.B.20 or 6.B.23.

Appendix 6.C shows that

$$\frac{\partial^2 L}{\partial(\sigma_i^2)^2} = -\frac{0.5nb_{1i}^2}{\left(b_{0i} + b_{1i}\sigma_i^2\right)^2} \tag{6.4.3}$$

$$\frac{\partial^2 L}{\partial(\sigma_\mu^2)^2} = -\frac{0.5nd_2^2}{\left(d_2\sigma_\mu^2 + 1\right)^2} \tag{6.4.4}$$

$$\frac{\partial L^2}{\partial\sigma_i^2 \, \partial\sigma_j^2} = \frac{n\beta_j^2\sigma_\mu^2\left(b_{1i}\sigma_i^2 + 0.5b_{0i}\right)}{\sigma_j^4\left(b_{0i} + b_{1i}\sigma_i^2\right)^2}$$

$$-\frac{0.5(n-1)\left(b_{0i}g_{3ij} - b_{1i}g_{2ij} - b_{2i}g_{1ij}\right)}{\left(b_{0i} + b_{1i}\sigma_i^2\right)^2} \tag{6.4.5}$$

$$\frac{\partial^2 L}{\partial\sigma_\mu^2 \, \partial\sigma_i^2} = \frac{n\beta_i^2\left(d_2\sigma_\mu^2 + 0.5\right) - (n-1)\beta_i\left(\beta_i s_i^2/\sigma_i^2 + \sum_{j\neq i}\beta_j s_{ij}/\sigma_j^2\right)}{\sigma_i^4\left(d_2\sigma_\mu^2 + 1\right)^2} \tag{6.4.6}$$

where b_{0i}, b_{1i}, and b_{2i} are given by Eqs. 6.B.16 to 6.B.18, d_2 by Eq. 6.B.25, and where

$$g_{1ij} = -\frac{\beta_j^2\sigma_\mu^2}{\sigma_j^4} \tag{6.4.7}$$

$$g_{2ij} = \frac{\sigma_\mu^2\left(-\beta_j^2 s_i^2 - \beta_i^2 s_j^2 + 2\beta_i\beta_j s_{ij}\right)}{\sigma_j^4} \tag{6.4.8}$$

$$g_{3ij} = \frac{-s_j^2}{\sigma_j^4} - \frac{\sigma_\mu^2}{\sigma_j^4}\left[\sum_{k\neq i,j}\frac{\left(\beta_k^2 s_j^2 + \beta_j^2 s_k^2 - 2\beta_j\beta_k s_{jk}\right)}{\sigma_k^2}\right] \tag{6.4.9}$$

In evaluating the second order derivatives, the parameter values are replaced by their estimates.

If a joint confidence region is desired, a procedure analogous to that described in Chapter 5 may be applied. Note that this entire development assumes that β_i is known for all i.

Table 6.20 Parameter Estimates for Example 4.G

i	$\tilde{\sigma}_i^2$	var $\tilde{\sigma}_i^2$	$df_i = 2\tilde{\sigma}_i^4/\text{var }\tilde{\sigma}_i^2$
1	0.000641	3.71×10^{-8}	22.1
2	0.000063	0.80×10^{-8}	1.0
3	0.000249	1.21×10^{-8}	10.2

EXAMPLE 6.M

In Example 6.C dealing with the fuse burning data, the parameter estimates for the various estimators were given in Table 6.4. As calculated in Example 4.G for the ME estimators derived from the paired differences, the sampling variances and approximate degrees of freedom were as shown in Table 6.20. The MLE's are $(n - 1)/n$ times the ME's, and these results apply to both estimators.

Consider the estimators based on the original data (OD), first for the ME's and then for the MLE's. Having excluded ME's for the constant bias model from further consideration (note from Table 6.6 that two of the three ME estimates were negative), attention is restricted to the nonconstant bias model. From Eq. 6.4.1,

$$\text{var }\tilde{\sigma}_1^2 = \frac{2}{28}\left\{(625)^2 + 0.5(1.0104)^4\left[\frac{(625)(63)}{(1.0104)^2(0.9985)^2}\right.\right.$$

$$\left.\left. + \frac{(625)(239)}{(1.0104)^2(0.9913)^2} + \frac{(63)(239)}{(0.9985)^2(0.9913)^2}\right]\right\} \times 10^{-12}$$

$$= 3.55 \times 10^{-8}$$

$$\text{var }\tilde{\sigma}_2^2 = 0.75 \times 10^{-8} \qquad \text{var }\tilde{\sigma}_3^2 = 1.11 \times 10^{-8}$$

The respective degrees of freedom are

$$df_1 = 22.0 \qquad df_2 = 1.1 \qquad df_3 = 10.3$$

in agreement with the estimates based on the paired differences. Note that Eqs. 4.4.1 and 6.4.1 are identical for $N = 3$ and for $\beta_i = 1$ for all i. More generally, Eqs. 5.4.1 and 6.4.1 are identical for all N when $\beta_i = 1$ for all i.

Table 6.21 Results for MLE's

Parameter	CB			NCB		
	Estimate	Var	df_i	Estimate	Var	df_i
σ_1^2	619×10^{-6}	3.29×10^{-8}	23.3	609×10^{-6}	3.17×10^{-8}	23.4
σ_2^2	62×10^{-6}	0.71×10^{-8}	1.1	64×10^{-6}	0.66×10^{-8}	1.2
σ_3^2	237×10^{-6}	1.08×10^{-8}	10.4	234×10^{-6}	0.99×10^{-8}	11.1
σ_μ^2	$43,586 \times 10^{-6}$	1.31×10^{-4}	29.0	$43,764 \times 10^{-6}$	1.32×10^{-4}	29.0

For the MLE's, the H matrix is constructed for each of the constant bias and nonconstant bias models. The results are as follows, applying Eqs. 6.4.2 to 6.4.9:

$$
\text{CB:} \quad H = \begin{pmatrix}
32,502,504 & & \text{(symmetric matrix)} & \\
19,079,346 & 266,968,040 & & \\
1,459,158 & 139,747,660 & 167,979,443 & \\
769.63 & -16,456.45 & -5,023.53 & 7,616.67
\end{pmatrix}
$$

$$
\text{NCB:} \quad H = \begin{pmatrix}
33,271,739 & & \text{(symmetric matrix)} & \\
20,784,949 & 267,370,954 & & \\
7,466,594 & 137,065,980 & 171,539,175 & \\
-222.73 & -19,764.96 & -1,451.64 & 7,554.58
\end{pmatrix}
$$

After the inverses are found, the sampling variances of the estimates and the respective degrees of freedom given in Table 6.21 can be calculated.

In this example, there is little basis for choosing among the various estimators. This is explained by the facts that β_i is nearly equal to unity for all i, and further, $N = 3$. Note in the examples of this chapter that for larger N, differences between the estimates are more pronounced, and there would be greater incentive to choose one estimator over the others. ■

6.4.2 Hypothesis Testing

As was true for the constant bias model, the hypothesis tests under consideration are large-sample likelihood-ratio tests that assume that the parameters are estimated by maximum likelihood. For the random model, the logarithm of the likelihood, denoted by L, is given in Appendix 6.B as Eq. 6.B.7 and is repeated here for reference:

$$
L = C - 0.5n \ln Q - 0.5(n-1) \sum_{i=1}^{N} \sum_{j=1}^{N} \sigma^{ij} s_{ij} \tag{6.4.10}
$$

where

$$Q = V\left(\sum_{j=1}^{N} \frac{a_j^2}{\sigma_j^2} + 1\right) \tag{6.4.11}$$

$$V = \prod_{j=1}^{N} \sigma_j^2 \tag{6.4.12}$$

$$\sigma^{ii} = \frac{V\left(\sum_{\substack{j=1 \\ j \neq i}}^{N} a_j^2/\sigma_j^2 + 1\right)}{Q\sigma_i^2} \tag{6.4.13}$$

$$\sigma^{ij} = \frac{-Va_i a_j}{Q\sigma_i^2 \sigma_j^2} \tag{6.4.14}$$

$$a_i = \beta_i \sigma_\mu \tag{6.4.15}$$

As a large-sample test of any given hypothesis about the parameters, the maximum value of L is found under that hypothesis and is denoted by $L(\hat{\hat{\omega}})$. The value of L corresponding to the MLE's of the parameters is also calculated and is denoted by $L(\hat{\Omega})$, given by

$$L(\hat{\Omega}) = C - 0.5n \ln \hat{Q} - 0.5(n-1) \sum_{i=1}^{N} \sum_{j=1}^{N} \sigma^{ij} s_{ij} \tag{6.4.16}$$

where $\hat{Q}$ is given by Eq. 6.4.11 and where σ^{ij} is evaluated using Eq. 6.4.13 or 6.4.14, with the parameters replaced by their MLE's in both instances.

Having found $L(\hat{\Omega})$ and $L(\hat{\hat{\omega}})$ for a specified hypothesis, large-sample theory tells us that under this specified hypothesis

$$\lambda = -2\left[L(\hat{\hat{\omega}}) - L(\hat{\Omega})\right] \tag{6.4.17}$$

is distributed approximately as chi-square with $(N - r)$ degrees of freedom, where r is the number of parameters estimated under the hypothesis. Thus the test statistic is λ, and the specified hypothesis is rejected if λ exceeds a prespecified critical value that is a function of the $(N - r)$ degrees of freedom and of α, the test significance level or type 1 error probability.

In calculating $L(\hat{\hat{\omega}})$ for a specified hypothesis, one hypothesis of interest is that the σ_i^2's are all equal. Letting σ_0^2 denote this common value so that the hypothesis is written:

$$H_0: \quad \sigma_1^2 = \sigma_2^2 = \cdots = \sigma_N^2 = \sigma_0^2 \tag{6.4.18}$$

then it is shown in Appendix 6.D that the MLE of σ_0^2 is the solution to the following cubic equation:

$$c_3\sigma_0^6 + c_2\sigma_0^4 + c_1\sigma_0^2 + c_0 = 0 \tag{6.4.19}$$

where

$$c_3 = nN \tag{6.4.20}$$

$$c_2 = nb_0(2N - 1) - (n - 1)b_1 \tag{6.4.21}$$

$$c_1 = n(N - 1)b_0^2 - 2(n - 1)(b_0b_1 - b_3 - 2b_2) \tag{6.4.22}$$

$$c_0 = (n - 1)b_0(2b_2 + b_3 - b_0b_1) \tag{6.4.23}$$

and where b_0, b_1, b_2, and b_3 are given by Eq. 6.D.5 and Eqs. 6.D.7 to 6.D.9.
   The test of the hypothesis indicated by Eq. 6.4.18 is illustrated by the following examples, as are tests of other hypotheses.

EXAMPLE 6.N

Example 6.L was concerned with the "measurement" of the lengths of 15 line segments by each of 6 individuals. Assuming a random model, test the hypothesis that the precisions are equal:

$$H_0: \quad \sigma_1^2 = \sigma_2^2 = \cdots = \sigma_6^2 = \sigma_0^2$$

The MLE's, given in Table 6.19, range from a low of 2.39 ($\hat{\sigma}_4^2$) units^2 to a high of 14.17 ($\hat{\sigma}_6^2$) units^2, the measurement unit being $\frac{1}{16}$ in. The ratio of $\hat{\sigma}_6^2$ to $\hat{\sigma}_4^2$ is almost 6. To test H_0, $L(\hat{\Omega})$ is calculated from Eq. 6.4.16 after having calculated $\hat{V}$ from Eq. 6.4.12 and $(\hat{Q})$ from Eq. 6.4.11.
   The values of s_i^2 and s_{ij} needed to perform the calculations are given in Table 6.22:

$$\hat{V} = (5.69)(3.38)\ldots(14.17) = 35{,}169$$

$$\hat{Q} = \hat{V}\left\{334.46\left[\frac{(0.9970)^2}{5.69} + \cdots + \frac{(0.9297)^2}{14.17}\right] + 1\right\}$$

$$= 431.00\hat{V} = 15{,}158{,}001$$

Table 6.22   Sample Variances and Covariances for Example 6.N

i	s_i^2	i, j	s_{ij}	i, j	s_{ij}
1	363.257143	1, 2	440.985714	2, 6	413.828571
2	549.266667	1, 3	352.414286	3, 4	311.457143
3	351.457143	1, 4	318.557143	3, 5	353.671429
4	286.885714	1, 5	361.342857	3, 6	328.685714
5	375.266667	1, 6	333.942857	4, 5	325.314286
6	327.314286	2, 3	434.885714	4, 6	297.828571
		2, 4	393.742857	5, 6	333.828571
		2, 5	445.861905		

Also,

$$-0.5(n - 1) \sum_{i=1}^{N} \sum_{j=1}^{N} \sigma^{ij} s_{ij} = -45.86$$

Then, from Eq. 6.4.16,

$$L(\hat{\Omega}) = C - 124.01 - 45.86$$

$$= C - 169.87$$

Next, σ_0^2 is estimated by solving the cubic Eq. 6.4.19. The b_i and c_i quantities are evaluated. From Eq. 6.D.5,

$$b_0 = 334.46 \sum_{j=1}^{6} \bar{\beta}_j^2 = 2052.69$$

From Eqs. 6.D.7 to 6.D.9,

$$b_1 = 2253.447620$$

$$b_2 = 1,869,802$$

$$b_3 = 809,778$$

From Eqs. 6.4.20 to 6.4.23,

$$c_3 = 90$$

$$c_2 = 307,146$$

$$c_1 = 313,880,291$$

$$c_0 = -2,191,171,713$$

With the above values for the coefficients of the cubic equation, the solution of Eq. 6.4.19 for σ_0^2 is

$$\hat{\sigma}_0^2 = 6.93 \text{ units}^2$$

Next, $L(\hat{\omega})$ is calculated from Eq. 6.D.10.

$$L(\hat{\omega}) = C - 72.59 - 57.23 - 45.04$$

$$= C - 174.86$$

The test statistic λ is

$$\lambda = -2(-174.86 + 169.87)$$

$$= 9.98$$

For a significance level $\alpha = 0.05$, and with $(6 - 1)$ or 5 degrees of freedom, the critical value is 11.07. Since $\lambda < 11.07$, the hypothesis of equal precisions is not rejected.   ∎

EXAMPLE 6.O

In the example just completed, and assuming a random model, test the fully specified hypothesis:

$$H_{01}: \quad \sigma_1 = \sigma_2 = \sigma_4 = 2 \text{ units}$$

$$\sigma_3 = \sigma_5 = 3 \text{ units}$$

$$\sigma_6 = 4 \text{ units}$$

The likelihood $L(\hat{\Omega})$ is, of course, unchanged and is $C - 169.87$. Under H_{01}, the likelihood is calculated following the same steps as in calculating $L(\hat{\Omega})$. The quantities V, Q, and so on, are marked with an asterisk to denote that they are calculated under H_{01}.

$$V^* = 82,944$$

In calculating Q^* from Eq. 6.4.11, it is first necessary to find the MLE of σ_μ^2 under the hypothesis. This quantity is calculated from Eq. 6.3.7 with d_2, d_3, and d_4 calculated from Eqs. 6.3.8 to 6.3.10 upon inputting the hypothesized values for $\sigma_1^2, \sigma_2^2, \ldots, \sigma_6^2$. From Eq. 6.3.8,

$$d_2 = \frac{(0.9970)^2}{4} + \cdots + \frac{(0.9297)^2}{16} = 1.0987$$

$$d_3 = \frac{(363.257143)}{4} + \cdots + \frac{(327.314286)}{16} = 401.0566$$

$$d_4 = \left[\frac{(0.9970)^2(549.266667)}{16} + \cdots + \frac{(0.9297)^2(375.266667)}{144} \right]$$

$$- \left[\frac{2(0.9970)(1.2321)(440.985714)}{16} + \cdots \right.$$

$$\left. + \frac{2(1.0058)(0.9297)(333.828571)}{144} \right]$$

$$= 341.8227 - 336.5410 = 5.2817$$

From Eq. 6.3.7,

$$\sigma_\mu^{*2} = 335.70$$

From Eq. 6.4.11,

$$Q^* = V^* \left\{ 335.70 \left[\frac{(0.9970)^2}{4} + \cdots + \frac{(0.9297)^2}{16} \right] + 1 \right\}$$

$$= 369.8327V^* = 30{,}675{,}407$$

The last set of terms in the expression for $L(\hat{\omega})$ are

$$-0.5(n-1) \sum_{i=1}^{6} \sum_{j=1}^{6} \sigma^{ij*} s_{ij}$$

where for $j = i$, σ^{ii*} is calculated from Eq. 6.4.13:

$$\sigma^{ii*} = \frac{335.70}{369.8327\sigma_i^2} \left(\sum_{\substack{j=1 \\ j \neq i}}^{6} \frac{\bar{\beta}_j^2}{\sigma_j^2} \right) + \frac{1}{369.8327\sigma_i^2}$$

where σ_i^2 and σ_j^2 are the hypothesized values. Also, σ^{ij*} is calculated from Eq. 6.4.14:

$$\sigma^{ij*} = - \frac{0.9077\bar{\beta}_i\bar{\beta}_j}{\sigma_i^2\sigma_j^2}$$

The σ^{ii*} values are calculated.

$$\sigma^{11*} = \frac{0.9077}{4}\left[\frac{(1.2321)^2}{4} + \cdots + \frac{(0.9297)^2}{16}\right] + \frac{1}{(369.8327)(4)} = 0.1936$$

$\sigma^{22*} = 0.1639 \qquad \sigma^{55*} = 0.0998$

$\sigma^{33*} = 0.1004 \qquad \sigma^{66*} = 0.0594$

$\sigma^{44*} = 0.2051$

Therefore,

$$\sum_{i=1}^{6}\sum_{j=1}^{6}\sigma^{ij*}s_{ij} = 311.3720 - 305.4783 = 5.8937$$

Then,

$$L(\omega^*) = C - 129.29 - 41.26 = C - 170.55$$

The test statistic is

$$\lambda^* = -2(-170.55 + 169.87) = 1.36$$

There are 6 degrees of freedom, and the $\alpha = 0.05$ significance level is 12.59. Since λ^* is less than 12.59, the hypothesis H_{01} is not rejected.   ∎

EXAMPLE 6.P

In Example 4.A, measurement error parameters were estimated for three methods of measuring reactor fuel pellet densities. The results were given in Table 6.4. Assuming a random model and using the MLE's for σ_i^2, $i = 1, 3, 6$, and for σ_μ^2, test the hypothesis that $\beta_1 = \beta_3 = \beta_6$ against the alternative that at least one of the inequalities is invalid.

We use the large-sample test based on the likelihood ratio to test this hypothesis, even though $\bar{\beta}_i$ is not an MLE. The test is an approximate one.

In $\hat{\Omega}$ space, $L(\hat{\Omega})$ is computed from Eq. 6.4.16 using the nonconstant bias MLE values to compute $\hat{Q}$. Under the hypothesis, $L(\hat{\omega})$ is also computed from Eq. 6.4.16, this time using the constant bias MLE values to compute $\hat{Q}$.

In $\hat{\Omega}$ space, from Eqs. 6.4.12 and 6.4.11,

$$\hat{V} = (3195)(26,246)(39,024) \times 10^{-18} = 3.2724 \times 10^{-6}$$

$$\hat{Q} = \hat{V}\left\{43,355\left[\frac{(1.0587)^2}{3195} + \frac{(1.0434)^2}{26,246} + \frac{(0.9053)^2}{39,024}\right] + 1\right\}$$

$$= 18.9184\hat{V} = 6.1908 \times 10^{-5}$$

and

$$-0.5n \ln \hat{Q} = 208.33$$

Also,

$$-0.5(n-1) \sum_{i=1}^{N} \sum_{j=1}^{N} \sigma^{ij}s_{ij} = -61.28$$

so that

$$L(\hat{\Omega}) = C + 208.33 - 61.28 = C + 147.05$$

Under the hypothesis,

$$\hat{V} = (4704)(25,135)(38,123) \times 10^{-18} = 4.5075 \times 10^{-6}$$

$$\hat{Q} = \hat{V}\left[49,088\left(\frac{1}{4704} + \frac{1}{25,135} + \frac{1}{38,123}\right) + 1\right]$$

$$= 14.6760\hat{V} = 6.6152 \times 10^{-5}$$

and $-0.5n \ln \hat{Q} = 206.91$ under the hypothesis. Also,

$$-0.5(n-1) \sum_{i=1}^{N} \sum_{j=1}^{N} \sigma^{ij}s_{ij} = -60.44$$

so that

$$L(\hat{\omega}) = C + 206.91 - 60.44 = C + 146.47$$

$$\lambda = -2(146.47 - 147.05) = 1.16$$

There are $(6 - 4) = 2$ degrees of freedom, and the $\alpha = 0.05$ critical value is 5.99. Since $\lambda < 5.99$, the hypothesis is not rejected; the constant bias model is judged to be appropriate. ∎

6.5 CONSTRAINED MAXIMUM LIKELIHOOD ESTIMATORS (CMLE'S)

This section treats the estimation problem in which $\sigma_i^2 = 0$ for some i. This may occur either because the MLE of one of the parameters is negative, in which case the estimate of that parameter is set equal to zero, or because one of the measurement methods is assumed to be without error. For example, a "method" could correspond to standard values assigned to the items being measured, which would be assumed to have negligibly small uncertainties.

6.5.1 Constrained Maximum Likelihood Estimates (CMLE's) for Random Model

In the iterative estimation process, one of the estimates, say that of σ_i^2, may become negative. If this happens, the iteration process should be continued to make sure that the final estimate of the parameter is in fact negative.

When $\hat{\sigma}_i^2$ is negative, then the estimate of σ_i^2 is zero, and the CMLE of σ_k^2, as derived in Appendix 6.E, is

$$\hat{\sigma}_k^2 = \frac{(n - 1)\left(a_i^2 s_k^2 + a_k^2 s_i^2 - 2a_i a_k s_{ik}\right)}{n a_i^2} \tag{6.5.1}$$

If there is doubt about which of the N parameters should be equated to zero, Appendix 6.E shows that it is one that minimizes

$$\sum_{k \neq i} \ln \hat{\sigma}_k^2 \qquad \text{or, equivalently,} \qquad \prod_{k \neq i} \hat{\sigma}_k^2$$

6.5.2 Constrained Maximum Likelihood Estimates (CMLE's) for Fixed Model

The model is rewritten with the constraint $\sigma_i^2 = 0$ imposed, and the CMLE's of the remaining parameters are found. The derivation is given in Appendix 6.F with the results shown here. The CMLE's of σ_j^2 for $j \neq i$ are

$$\hat{\sigma}_j^2 = \frac{\left(s_j^2 - s_{ij}^2/s_i^2\right)(n - 1)}{n} \tag{6.5.2}$$

Since $\sigma_i^2 = 0$, it is logical to set $\beta_i = 1$ and $\alpha_i = 0$, in which case the CMLE's of β_j and α_j are, for $j \neq i$,

$$\hat{\beta}_j = \frac{s_{ij}}{s_i^2} \tag{6.5.3}$$

$$\hat{\alpha}_j = \bar{x}_j - \hat{\beta}_j \bar{x}_i \tag{6.5.4}$$

Keep in mind that $\hat{\beta}_j$ is really the estimate of β_j/β_i, and $\hat{\alpha}_j$ is the estimate of $(\alpha_j - \alpha_i)$. To continue, under the constraints imposed,

$$\hat{\mu}_k = x_{ik} \qquad \text{for } k = 1, 2, \ldots, n \tag{6.5.5}$$

although these estimates are not too useful since only the relative biases may be estimated.

6.5.3   Interval Estimation for Constrained Maximum Likelihood Estimates (CMLE's)

In the event $\sigma_i^2 = 0$ for some i, then the expression for $\partial L / \partial \sigma_k^2$, $k \neq i$, is given by Eq. 6.E.3. The second partial derivative is

$$\frac{\partial^2 L}{\partial \left(\sigma_k^2 \right)^2} = \frac{0.5n}{\sigma_k^4} - \frac{(n-1)}{\sigma_k^6} \left[\frac{a_k^2 s_i^2}{a_i^2} + s_k^2 - \frac{2 a_k s_{ik}}{a_i} \right] \tag{6.5.6}$$

Also,

$$\frac{\partial^2 L}{\partial \left(\sigma_k^2 \sigma_j^2 \right)} = 0 \tag{6.5.7}$$

In evaluating Eq. 6.5.6 for $\hat{\sigma}_k^2$ given by Eq. 6.5.1, note that the expression in the square brackets is simply $n\hat{\sigma}_k^2/(n-1)$ so that

$$\frac{\partial^2 L}{\partial \left(\sigma_k^2 \right)^2} = -\frac{0.5n}{\hat{\sigma}_k^4} \tag{6.5.8}$$

Thus

$$\text{var } \hat{\sigma}_k^2 = \frac{2\hat{\sigma}_k^4}{n} \tag{6.5.9}$$

and

$$df_k = \frac{2n\hat{\sigma}_k^4}{2\hat{\sigma}_k^4} = n \tag{6.5.10}$$

Thus as would be expected intuitively, if $\sigma_i^2 = 0$, then there are n degrees of freedom for the estimates of σ_k^2, $k \neq i$.

For the fixed model, the expression for $\partial L / \partial \sigma_j^2$, $j \neq i$, is given by Eqs. 6.F.7 and 6.F.9. The second partial derivative is

$$\frac{\partial^2 L}{\partial \left(\sigma_j^2 \right)^2} = \frac{0.5n}{\sigma_j^4} - \frac{(n-1)\left(s_j^2 - s_{ij}^2 / s_i^2 \right)}{\sigma_j^6} \tag{6.5.11}$$

and, as for the random model, there are n degrees of freedom associated with the estimate of σ_j^2 given by Eq. 6.5.2.

EXAMPLE 6.Q

In Example 6.N dealing with automobile fuel consumption, the estimate of σ_2^2 was negative for both the constant bias and nonconstant bias models (see Table 6.8). Setting $\hat{\sigma}_2^2 = 0$ and applying Eq. 6.5.1, the constant bias model,

$$\hat{\sigma}_1^2 = \frac{15(21.598460 + 25.779234 - 47.092662)}{16}$$

$$= 0.2672$$

$$\hat{\sigma}_3^2 = 0.0371$$

For the nonconstant bias model,

$$\hat{\sigma}_1^2 = \frac{15}{16} \left[(1.0264)^2 (21.598460) + (0.9373)^2 (25.779234) \right.$$

$$\left. - 2(1.0264)(0.9373)(23.546331) \right]$$

$$= 0.0905$$

$$\hat{\sigma}_3^2 = 0.0349 \qquad\qquad \blacksquare$$

EXAMPLE 6.R

In Example 6.L dealing with the lengths of line segments, assume that $\sigma_7^2 = 0$, with Method 7 corresponding to the actual measured segment lengths. Assuming a fixed model, the CMLE's of the remaining parameters are given by Eq. 6.5.2. The relevant sample variances and covariances calculated from the Table

6.18 data are as follows:

$$s_1^2 = 363.257143 \qquad s_{17} = 375.628571$$

$$s_2^2 = 549.266667 \qquad s_{27} = 464.671429$$

$$s_3^2 = 351.457143 \qquad s_{37} = 370.171429$$

$$s_4^2 = 286.885714 \qquad s_{47} = 334.957143$$

$$s_5^2 = 375.266667 \qquad s_{57} = 380.957143$$

$$s_6^2 = 327.314286 \qquad s_{67} = 352.114286$$

$$s_7^2 = 396.600000$$

Then

$$\hat{\sigma}_1^2 = \frac{14}{15}\left[363.257143 - \frac{(375.628571)^2}{396.6}\right] = 6.99$$

$$\hat{\sigma}_2^2 = 4.52$$

$$\hat{\sigma}_3^2 = 5.56 \qquad \hat{\sigma}_5^2 = 8.71$$

$$\hat{\sigma}_4^2 = 3.72 \qquad \hat{\sigma}_6^2 = 13.72 \qquad\qquad ■$$

6.6 ESTIMATION OF RELATIVE BIASES

For the general model, the relative bias of a given method involves both α_i and β_i. Under the constraint that the geometric mean of the β's be 1, the estimate of β_i was given by Eq. 6.2.4. Similarly, applying the constraint that the α's sum to zero, α_i is estimated by

$$\bar{\alpha}_i = \bar{x}_{i\bullet} - \bar{\beta}_i \bar{x}_{\bullet\bullet} \qquad\qquad (6.6.1)$$

where

$$\bar{x}_{i\bullet} = \frac{\sum\limits_{k=1}^{n} x_{ik}}{n} \qquad\qquad (6.6.2)$$

and

$$\bar{x}_{\cdot\cdot} = \frac{\sum\limits_{i=1}^{N} \bar{x}_{i\cdot}}{N} \qquad (6.6.3)$$

For $\beta_i \neq 1$ for all i, a result parallel to Eq. 5.6.1 does not exist except at a fixed value of μ_i. If $\beta_i = 1$ for all i, then

$$\bar{\alpha}_i - \bar{\alpha}_j = \bar{x}_{i\cdot} - \bar{x}_{j\cdot} \qquad (6.6.4)$$

which is equivalent to Eq. 4.7.1.

6.7 GUIDANCE IN CHOOSING AN ESTIMATOR

In this chapter and in Chapters 4 and 5, a number of estimators of σ_i^2 are given, suggesting a problem of which estimator should be used in practice. There is no right or wrong answer. In a general sense, some estimators may be attractive because of their simplicity and others because they possess better statistical properties as estimators. Ultimately, the choice becomes a matter of personal preference.

Nevertheless, some guiding comments may be helpful. If there is a priori reason to believe that the constant bias model applies, for example, based on prior experience, parameter estimation based on the paired differences (Chapters 4 and 5) has the advantage that one need make no assumptions about μ_k, that is, about whether the model is random or fixed. The ME (Grubbs') estimates of σ_i^2 are found rather simply and, for $N = 3$, are essentially equivalent to the MLE's. However, for $N \geq 4$, experience has shown that the likelihood of the ME's may be quite far removed from the optimum set of estimates, and the MLE's are preferred. Although the computational effort in finding MLE's is considerable if done by hand, the methods of Chapter 5 are readily programmable; in fact, a Fortran listing of an MLE estimation program is given in Chapter 9. ME's may be used to provide quick, relatively simple estimates if computer assistance is not readily available.

Even when the constant bias model may be assumed, the estimators based on the original data (Chapter 6) are preferred over those based on paired differences if inferences are also to be made about σ_μ^2, the process variance. In this case, the ME's are generally of little value, and the MLE's are required. The ME's provide acceptable estimates only if σ_μ^2 is very small in a relative sense. When obtaining MLE's based on original data as compared to those based on paired differences, one tends to sacrifice efficiency in estimating σ_i^2 for the ability to estimate σ_μ^2. For paired differences, the MLE of σ_i^2 is independent of σ_μ^2, but such is not the case for estimates based on the original data.

Finally, if evidence suggests that the nonconstant bias model may be appropriate, then the MLE's of Chapter 6 for this model are the ones to use, again utilizing the Fortran program listing in Chapter 9 if computer support is available. Otherwise, the simpler ME's should provide reasonably good estimates of the parameters in most instances. If one is uncertain whether the constant or nonconstant bias model is valid, the first step could be to test the hypothesis: $\beta_1 = \beta_2 = \cdots = \beta_N$, as was done in Example 6.P, and then use paired differences if the hypothesis is not rejected.

LEAST SQUARES ESTIMATORS (LSE'S) DERIVED FROM MOMENT EQUATIONS

There are $N(N+1)/2$ moment equations:

$$s_i^2 = \beta_i^2 \sigma_\mu^2 + \sigma_i^2; \qquad i = 1, 2, \ldots, N \tag{6.A.1}$$

and

$$s_{ij} = \beta_i \beta_j \sigma_\mu^2; \qquad i = 1, 2, \ldots N - 1, \ j > i \tag{6.A.2}$$

It is not possible to estimate β_i unless one or more of the β_j's is known, or, in the case of the fixed model, unless the μ_k's are given. It is possible, however, to estimate relative biases by imposing some constraint on the parameters. A reasonable constraint is that the geometric mean of the β_i's be unity:

$$\prod_{i=1}^{N} \beta_i = 1 \tag{6.A.3}$$

The parameters σ_μ^2 and β_i are estimated from Eq. 6.A.2, and these estimates are then used to estimate σ_i^2 from Eq. 6.A.1, wherein it follows immediately that

$$\tilde{\sigma}_i^2 = s_i^2 - \bar{\beta}_i^2 \tilde{\sigma}_\mu^2 \tag{6.A.4}$$

where $\bar{\beta}_i^2$ is given by Eq. 6.A.16 and $\tilde{\sigma}^2$ by Eq. 6.A.17.

In operations on Eqs. 6.A.2, they and the constraint Eq. 6.A.3 are written in their logarithmic equivalent forms:

$$\sum_{i=1}^{N} \ln \beta_i = 0 \tag{6.A.5}$$

$$\ln s_{ij} = \ln \beta_i + \ln \beta_j + \ln \sigma_\mu^2 \tag{6.A.6}$$

The $N(N-1)/2$ moment equations, Eq. 6.A.6, are solved by least squares upon writing the constraint, Eq. 6.A.5, in its equivalent form:

$$\ln \beta_1 = - \sum_{j=2}^{N} \ln \beta_j \qquad (6.A.7)$$

The least squares estimates of $\beta_2, \beta_3, \ldots, \beta_N, \sigma_\mu^2$ are the solutions of the matrix equation

$$(\bar{\beta}) = X^{-1}(s) \qquad (6.A.8)$$

where $(\bar{\beta})$ is the column matrix with N rows with element $\ln \bar{\sigma}_\mu^2$ in row 1 and element $\ln \bar{\beta}_i$ in row i, $i > 1$, and where (s) is the column matrix with N rows. The element of (s) in row 1 is

$$\sum_{i=1}^{N-1} \sum_{j>i} \ln s_{ij} \qquad (6.A.9)$$

and the element in row i for $i > 1$ is

$$- \sum_{\substack{j=2 \\ j\neq i}}^{N} \ln s_{1j} + \sum_{j=2}^{N} \ln s_{ij} \qquad (6.A.10)$$

The elements of the symmetric $(N + 1)$ square X matrix are as follows:

$$\text{element } (1,1) = \frac{N(N+1)}{2} \qquad (6.A.11)$$

$$\text{element } (1, j) = 0; \qquad j = 2, 3, \ldots, N \qquad (6.A.12)$$

$$\text{element } (i, i) = 2(N-2); \qquad i = 2, 3, \ldots, N \qquad (6.A.13)$$

$$\text{element } (i, j) = (N-2); \qquad i = 2, 3, \ldots, N-1; \; j > i \qquad (6.A.14)$$

Equations 6.A.11 and 6.A.12 indicate that the estimate of β_i is independent of the estimate of σ_μ^2 for all i. The inverse of X has element $2/N(N-1)$ in position $(1,1)$ and 0 elsewhere in row 1. The element in position (i, i) is $(N-1)/N(N-2)$ for $i > 1$, and the off-diagonal element is $-1/N(N-2)$ for $i > 1$ and $j \neq i$.

With $(\bar{\beta})$, X^{-1}, and (s) defined above, it follows from Eq. 6.A.6 that the estimate of $\ln \beta_i$ for $i > 1$ is

$$
\ln \bar{\beta}_i = \frac{(N-1)\left(-\sum\limits_{\substack{j=2 \\ j\neq i}}^{N} \ln s_{1j} + \sum\limits_{j=2}^{N} \ln s_{ij}\right) - \sum\limits_{\substack{k=2 \\ k\neq i}}^{N}\left(-\sum\limits_{\substack{j=2 \\ j\neq k}}^{N} \ln s_{1j} + \sum\limits_{j=2}^{N} \ln s_{kj}\right)}{N(N-2)}
$$

$$
= \frac{(N-2)\sum\limits_{\substack{j=1 \\ j\neq i}}^{N} \ln s_{ij} - 2\sum\limits_{\substack{j=1 \\ j\neq i}}^{N}\sum\limits_{\substack{k>j \\ k\neq i}}^{N} \ln s_{jk}}{N(N-2)} \tag{6.A.15}
$$

From Eq. 6.A.15, the estimate of β_i is

$$
\bar{\beta}_i = \frac{\left(\prod\limits_{j\neq i} s_{ij}\right)^{1/N}}{\left(\prod\limits_{\substack{j,k>j \\ \neq i}} s_{ij}\right)^{2/N(N-2)}} \tag{6.A.16}
$$

and from Eq. 6.A.7, Eq. 6.A.16 applies also for $i = 1$.

From the elements of X^{-1} and (s), the moment estimator of $\ln \sigma_\mu^2$ follows immediately:

$$
\ln \tilde{\sigma}_\mu^2 = \frac{2\sum\limits_{i=1}^{N-1}\sum\limits_{j>i} \ln s_{ij}}{N(N-1)} \tag{6.A.17}
$$

from which

$$
\tilde{\sigma}_\mu^2 = \prod\limits_{i,j>i} s_{ij}^{2/N(N-1)} \tag{6.A.18}
$$

MAXIMUM LIKELIHOOD ESTIMATORS (MLE'S) FOR RANDOM MODEL

The constraint is placed on the model that $\prod_{i=1}^{N} \beta_i = 1$. The estimates of the β's are given by Eq. 6.2.4. The MLE's of σ_i^2, $i = 1, 2, \ldots, N$, and of σ_μ^2 are then found from this starting point. First, find the MLE of σ_i^2.

For convenience in notation, replace $\beta_i \sigma_\mu$ by a_i. Then the variance-covariance matrix (σ_{ij}) is a symmetric matrix with diagonal element

$$a_i^2 + \sigma_i^2 \tag{6.B.1}$$

in position (i, i) and off-diagonal element

$$a_i a_j \tag{6.B.2}$$

in position (i, j) for $j \neq i$.

It can be shown that the determinant of (σ_{ij}), denoted by Q, is

$$Q = V\left(\sum_{j=1}^{N} \frac{a_j^2}{\sigma_j^2} + 1 \right) \tag{6.B.3}$$

where

$$V = \prod_{j=1}^{N} \sigma_j^2 \tag{6.B.4}$$

It can further be shown that the diagonal element in position (i, i) of the inverse matrix is

$$\sigma^{ii} = \frac{V\left(\sum_{\substack{j=1 \\ j \neq i}}^{N} a_j^2/\sigma_j^2 + 1 \right)}{Q \sigma_i^2} \tag{6.B.5}$$

and the off-diagonal element in position (i, j) is

$$\sigma^{ij} = -\frac{V a_i a_j}{Q \sigma_i^2 \sigma_j^2} \tag{6.B.6}$$

221

The logarithm of the likelihood for n observations is of the form

$$L = C - 0.5n \ln Q - 0.5(n-1) \sum_{i=1}^{N} \sum_{j=1}^{N} \sigma^{ij} s_{ij} \qquad (6.B.7)$$

where for $i = j$, s_{ii} is the same as s_i^2 of Eq. 1.6.2.

The expression for L is now rewritten by using Eqs. 6.B.5 and 6.B.6, and further, by isolating terms in σ_i^2 to facilitate partially differentiating L with respect to σ_i^2. L then becomes

$$L = C - 0.5n \left[\ln \sigma_i^2 + \sum_{j \neq i} \ln \sigma_j^2 + \ln \left(\frac{a_i^2}{\sigma_i^2} + \sum_{j \neq i} \frac{a_j^2}{\sigma_j^2} + 1 \right) \right]$$

$$- \frac{0.5(n-1)}{a_i^2/\sigma_i^2 + \sum_{j \neq i} a_j^2/\sigma_j^2 + 1}$$

$$\times \left[\frac{\left(\sum_{j \neq i} a_j^2/\sigma_j^2 + 1 \right) s_i^2}{\sigma_i^2} + (a_i^2/\sigma_i^2) \sum_{j \neq i} \frac{s_j^2}{\sigma_j^2} + \sum_{j \neq i} \frac{\left(\sum_{k \neq i, j} a_k^2/\sigma_k^2 + 1 \right) s_j^2}{\sigma_j^2} \right.$$

$$\left. - 2(a_i/\sigma_i^2) \sum_{j \neq i} \frac{a_j s_{ij}}{\sigma_j^2} - 2 \sum_{j \neq i} \sum_{\substack{k > j \\ \neq i}} \frac{a_j a_k s_{jk}}{\sigma_j^2 \sigma_k^2} \right] \qquad (6.B.8)$$

This expression is rewritten upon making the following substitutions:

$$c_{0i} = a_i^2 \qquad (6.B.9)$$

$$c_{1i} = \sum_{j \neq i} \frac{a_j^2}{\sigma_j^2} + 1 \qquad (6.B.10)$$

$$c_{2i} = \sum_{j \neq i} \frac{s_j^2}{\sigma_j^2} \qquad (6.B.11)$$

$$c_{3i} = \sum_{j \neq i} \frac{\left(\sum_{k \neq j, i} a_k^2/\sigma_k^2 + 1 \right) s_j^2}{\sigma_j^2}$$

$$= \sum_{j \neq i} \frac{\left(c_{1i} - a_j^2/\sigma_j^2 \right) s_j^2}{\sigma_j^2} \qquad (6.B.12)$$

$$c_{4i} = \sum_{j \neq i} \frac{a_j s_{ij}}{\sigma_j^2} \qquad (6.B.13)$$

$$c_{5i} = \sum_{j \neq i} \sum_{\substack{k > j \\ \neq i}} \frac{a_j a_k s_{jk}}{\sigma_j^2 \sigma_k^2} \qquad (6.B.14)$$

Then,

$$L = C - 0.5n \ln \sigma_i^2 - 0.5n \sum_{j \neq i} \ln \sigma_j^2 - 0.5n \ln\left(c_{0i} + c_{1i}\sigma_i^2 - \sigma_i^2 \right)$$

$$- \frac{0.5(n-1)\sigma_i^2}{c_{0i} + c_{1i}\sigma_i^2} \left(\frac{c_{1i} s_i^2}{\sigma_i^2} + \frac{c_{0i} c_{2i}}{\sigma_i^2} + c_{3i} - \frac{2a_i c_{4i}}{\sigma_i^2} - 2c_{5i} \right)$$

$$= C - 0.5n \sum_{j \neq i} \ln \sigma_j^2 - 0.5n \ln\left(c_{0i} + c_{1i}\sigma_i^2 \right)$$

$$- \frac{0.5(n-1)}{c_{0i} + c_{1i}\sigma_i^2} \left(c_{1i} s_i^2 + c_{0i} c_{2i} + c_{3i}\sigma_i^2 - 2a_i c_{4i} - 2c_{5i}\sigma_i^2 \right) \qquad (6.B.15)$$

Make the further simplifications in notation:

$$b_{0i} = c_{0i} \qquad (6.B.16)$$

$$b_{1i} = c_{1i} \qquad (6.B.17)$$

$$b_{2i} = c_{1i} s_i^2 + c_{0i} c_{2i} - 2a_i c_{4i} \qquad (6.B.18)$$

$$b_{3i} = c_{3i} - 2c_{5i} \qquad (6.B.19)$$

Then L may be written:

$$L = C - 0.5n \sum_{j \neq i} \ln \sigma_j^2 - 0.5n \ln\left(b_{0i} + b_{1i}\sigma_i^2 \right)$$

$$- \frac{0.5(n-1)}{\left(b_{0i} + b_{1i}\sigma_i^2 \right)} \left(b_{2i} + b_{3i}\sigma_i^2 \right) \qquad (6.B.20)$$

L is then differentiated with respect to σ_i^2 and equated to zero. Note that none of the b's are functions of σ_i^2.

$$\frac{\partial L}{\partial \sigma_i^2} = -\frac{0.5 n b_{1i}}{b_{0i} + b_{1i}\sigma_i^2} - 0.5(n-1)\frac{(b_{0i}b_{3i} - b_{1i}b_{2i})}{(b_{0i} + b_{1i}\sigma_i^2)^2} = 0 \quad (6.B.21)$$

This is easily solved for σ_i^2 to give the MLE:

$$\hat{\sigma}_i^2 = \frac{(n-1)(b_{1i}b_{2i} - b_{0i}b_{3i})}{n b_{1i}^2} - \frac{b_{0i}}{b_{1i}} \quad (6.B.22)$$

This is Eq. 6.3.1. The b's are defined by Eqs. 6.B.9 to 6.B.14 and 6.B.16 to 6.B.19 or, equivalently, by Eqs. 6.3.3 to 6.3.6 in the text.

Next, consider the MLE of σ_μ^2. Rewrite L of Eq. 6.B.8 as follows, after replacing a_i by $\beta_i\sigma_\mu$:

$$L = C - 0.5n\sum_i \ln\sigma_i^2 - 0.5n\ln\left(\sum_i \frac{\beta_i^2\sigma_\mu^2}{\sigma_i^2} + 1\right) - \frac{0.5(n-1)}{\sum_i \beta_i^2\sigma_\mu^2/\sigma_i^2 + 1}$$

$$\times \left(\sum_i \frac{s_i^2}{\sigma_i^2} + \sigma_\mu^2\sum_i\sum_{j\neq i} \frac{\beta_i^2 s_j^2}{\sigma_i^2\sigma_j^2} - 2\sigma_\mu^2\sum_i\sum_{j>i} \frac{\beta_i\beta_j s_{ij}}{\sigma_i^2\sigma_j^2}\right) \quad (6.B.23)$$

Make the following replacements:

$$d_1 = \sum_i \ln\sigma_i^2 \quad (6.B.24)$$

$$d_2 = \sum_i \frac{\beta_i^2}{\sigma_i^2} \quad (6.B.25)$$

$$d_3 = \sum_i \frac{s_i^2}{\sigma_i^2} \quad (6.B.26)$$

$$d_4 = \sum_i\sum_{j\neq i} \frac{\beta_i^2 s_j^2}{\sigma_i^2\sigma_j^2} - 2\sum_i\sum_{j>i} \frac{\beta_i\beta_j s_{ij}}{\sigma_i^2\sigma_j^2} \quad (6.B.27)$$

Then

$$L = C - 0.5n \ln\left(d_2\sigma_\mu^2 + 1\right) - \frac{0.5(n-1)}{d_2\sigma_\mu^2 + 1}\left(d_3 + d_4\sigma_\mu^2\right) \quad (6.B.28)$$

$$-\frac{\partial(2L)}{\partial\sigma_\mu^2} = \frac{nd_2}{d_2\sigma_\mu^2 + 1} + \frac{(n-1)(d_4 - d_2d_3)}{\left(d_2\sigma_\mu^2 + 1\right)^2} = 0 \quad (6.B.29)$$

from which

$$\hat{\sigma}_\mu^2 = \frac{(n-1)(d_2d_3 - d_4)}{nd_2^2} - \frac{1}{d_2} \quad (6.B.30)$$

which is Eq. 6.3.7 in the text.

DERIVATION OF ELEMENTS OF *H* MATRIX, SECTION 6.4.1

For simplicity in notation, replace σ_i^2 by v_i and σ_μ^2 by v_0. In deriving the elements of H, some intermediate results are needed, as follows. From Eqs. 6.B.16 to 6.B.19,

$$\frac{\partial b_{ki}}{\partial v_i} = 0; \qquad k = 0, 1, 2, 3 \tag{6.C.1}$$

$$\frac{\partial b_{0i}}{\partial v_j} = 0 \tag{6.C.2}$$

$$\frac{\partial b_{1i}}{\partial v_j} = -\frac{\beta_j^2 v_0}{v_j^2} \tag{6.C.3}$$

$$\frac{\partial b_{2i}}{\partial v_j} = \frac{v_0}{v_j^2}\left(-\beta_j^2 s_i^2 - \beta_i^2 s_j^2 + 2\beta_i\beta_j s_{ij}\right) \tag{6.C.4}$$

$$\frac{\partial b_{3i}}{\partial v_j} = -\frac{s_j^2}{v_j^2} - \frac{v_0}{v_j^2}\left[\sum_{k \neq i,j} \frac{\beta_k^2 s_j^2 + \beta_j^2 s_k^2 - 2\beta_j\beta_k s_{jk}}{v_k}\right] \tag{6.C.5}$$

Again to simplify notation, we henceforth denote $(\partial b_{ki}/\partial v_j)$ by g_{kij}. Continuing with intermediate results, from Eqs. 6.B.25 to 6.B.27,

$$\frac{\partial d_k}{\partial v_0} = 0; \qquad k = 1, 2, 3, 4 \tag{6.C.6}$$

$$\frac{\partial d_2}{\partial v_i} = -\frac{\beta_i^2}{v_i^2} \tag{6.C.7}$$

$$\frac{\partial d_3}{\partial v_i} = -\frac{s_i^2}{v_i^2} \tag{6.C.8}$$

$$\frac{\partial d_4}{\partial v_i} = -\frac{\beta_i^2}{v_i^2}\sum_{j\neq i}\frac{s_j^2}{v_j} - \frac{s_i^2}{v_i^2}\sum_{j\neq i}\frac{\beta_j^2}{v_j} + \frac{2\beta_i}{v_i^2}\sum_{j\neq i}\frac{\beta_j s_{ij}}{v_j}$$

$$= (1/v_i^2)\sum_{j\neq i}\frac{-\beta_i^2 s_j^2 - \beta_j^2 s_i^2 + 2\beta_i\beta_j s_{ij}}{v_j} \tag{6.C.9}$$

With these intermediate results, the elements of H are now found.

6.C.1 Derive Equation 6.4.3

From Eq. 6.B.21,

$$\frac{\partial L}{\partial v_i} = -\frac{0.5nb_{1i}}{b_{0i} + b_{1i}v_i} - \frac{0.5(n-1)(b_{0i}b_{3i} - b_{1i}b_{2i})}{(b_{0i} + b_{1i}v_i)^2} \tag{6.C.10}$$

$$\frac{\partial^2 L}{\partial v_i^2} = \frac{0.5nb_{1i}^2}{(b_{0i} + b_{1i}v_i)^2} + \frac{(n-1)(b_{0i}b_{3i} - b_{1i}b_{2i})b_{1i}}{(b_{0i} + b_{1i}v_i)^3} \tag{6.C.11}$$

From Eq. 6.C.10,

$$\frac{(n-1)(b_{0i}b_{3i} - b_{1i}b_{2i})b_{1i}}{(b_{0i} + b_{1i}v_i)^3} = -\frac{nb_{1i}^2}{(b_{0i} + b_{1i}v_i)^2} \tag{6.C.12}$$

and hence,

$$\frac{\partial^2 L}{\partial v_i^2} = -\frac{0.5nb_{1i}^2}{(b_{0i} + b_{1i}v_i)^2} \tag{6.C.13}$$

which is Eq. 6.4.3.

6.C.2 Derive Equation 6.4.4

From Eq. 6.B.29,

$$\frac{\partial L}{\partial v_0} = -\frac{0.5nd_2}{d_2v_0 + 1} - \frac{0.5(n-1)(d_4 - d_2d_3)}{(d_2v_0 + 1)^2} \tag{6.C.14}$$

$$\frac{\partial^2 L}{\partial v_0^2} = \frac{0.5nd_2^2}{(d_2v_0 + 1)^2} + \frac{(n-1)(d_4 - d_2d_3)d_2}{(d_2v_0 + 1)^3} \tag{6.C.15}$$

From Eq. 6.C.14,

$$\frac{(n-1)(d_4 - d_2 d_3)d_2}{(d_2 v_0 + 1)^3} = -\frac{nd_2^2}{(d_2 v_0 + 1)^2} \qquad (6.C.16)$$

and hence,

$$\frac{\partial^2 L}{\partial v_0^2} = -\frac{0.5 n d_2^2}{(d_2 v_0 + 1)^2} \qquad (6.C.17)$$

which is Eq. 6.4.4.

6.C.3   Derive Equation 6.4.5

From Eq. 6.C.10,

$$\frac{\partial^2 L}{\partial v_i \, \partial v_j} = -\frac{0.5n \left[(b_{0i} + b_{1i} v_i) g_{1ij} - b_{1i} v_i g_{1ij} \right]}{(b_{0i} + b_{1i} v_i)^2}$$

$$-\frac{0.5(n-1)}{(b_{0i} + b_{1i} v_i)^4} \left[(b_{0i} + b_{1i} v_0)^2 (b_{0i} g_{3ij} - b_{1i} g_{2ij} - b_{2i} g_{1ij}) \right.$$

$$\left. -2(b_{0i} b_{3i} - b_{1i} b_{2i})(b_{0i} + b_{1i} v_i) v_i g_{1ij} \right]$$

$$= -\frac{0.5 n b_{0i} g_{1ij}}{(b_{0i} + b_{1i} v_i)^2} - \frac{0.5(n-1)(b_{0i} g_{3ij} - b_{1i} g_{2ij} - b_{2i} g_{1ij})}{(b_{0i} + b_{1i} v_i)^2}$$

$$+ \frac{(n-1) v_i g_{1ij}(b_{0i} b_{3i} - b_{1i} b_{2i})}{(b_{0i} + b_{1i} v_i)^3}$$

The first and last terms are combined using Eq. 6.C.10, and g_{1ij} is replaced by its equivalent given by Eq. 6.C.3. The result is

$$\frac{\partial^2 L}{\partial v_i \, \partial v_j} = \frac{n \beta_j^2 v_0 (b_{1i} v_i + 0.5 b_{0i})}{v_j^2 (b_{0i} + b_{1i} v_i)^2}$$

$$-\frac{0.5(n-1)(b_{0i} g_{3ij} - b_{1i} g_{2ij} - b_{2i} g_{1ij})}{(b_{0i} + b_{1i} v_i)^2} \qquad (6.C.18)$$

which is Eq. 6.4.5.

6.C.4   Derive Equation 6.4.6

From Eqs. 6.C.14 and 6.C.7 to 6.C.9,

$$\frac{\partial^2 L}{\partial v_0 \, \partial v_i} = \frac{0.5 n \beta_i^2}{v_i^2 (d_2 v_{0i} + 1)^2} - \frac{0.5(n-1)}{(d_2 v_0 + 1)^4}$$

$$\times \left[(d_2 v_0 + 1)^2 \left(\frac{\partial d_4}{\partial v_i} - d_2 \frac{\partial d_3}{\partial v_i} - d_3 \frac{\partial d_2}{\partial v_i} \right) \right.$$

$$\left. + \frac{2(d_4 - d_2 d_3)(d_2 v_0 + 1) v_0 \beta_i^2}{v_i^2} \right]$$

consider the expression

$$\frac{\partial d_4}{\partial v_i} - d_2 \frac{\partial d_3}{\partial v_i} - d_3 \frac{\partial d_2}{\partial v_i}$$

$$= \left(\frac{1}{v_i^2} \right) \sum_{j \neq i} \frac{\left(-\beta_i^2 s_j^2 - \beta_j^2 s_i^2 + 2\beta_i \beta_j s_{ij} \right)}{v_j} + \frac{s_i^2}{v_i^2} \sum_j \frac{\beta_j^2}{v_j} + \frac{\beta_i^2}{v_i^2} \sum_j \frac{s_j^2}{v_j}$$

$$= \frac{1}{v_i^2} \sum_{j \neq i} \frac{\left(-\beta_i^2 s_j^2 - \beta_j^2 s_i^2 + 2\beta_i \beta_j s_{ij} + \beta_j^2 s_i^2 + \beta_i^2 s_j^2 \right)}{v_j} + \frac{2\beta_i^2 s_i^2}{v_i^3}$$

$$= \frac{2\beta_i^2 s_i^2}{v_i^3} + \left(\frac{2\beta_i}{v_i^2} \right) \sum_{j \neq i} \frac{\beta_j s_{ij}}{v_j}$$

Also, from Eq. 6.C.14, note that

$$\frac{(n-1)(d_4 - d_2 d_3) \beta_i^2 v_0}{v_i^2 (d_2 v_0 + 1)^3} = \frac{-n \beta_i^2 d_2 v_0}{v_i^2 (d_2 v_0 + 1)^2}$$

Hence, the expression for $(\partial^2 L / \partial v_0 \, \partial v_i)$ reduces to

$$\frac{\partial^2 L}{\partial v_0 \, \partial v_i} = \frac{n \beta_i^2 (d_2 v_0 + 0.5) - (n-1) \beta_i \left(\beta_i s_i^2 / v_i + \sum_{j \neq i} \beta_j s_{ij} / v_j \right)}{v_i^2 (d_2 v_0 + 1)^2}$$

$$(6.C.19)$$

which is Eq. 6.4.6.

DERIVATION OF MAXIMUM LIKELIHOOD ESTIMATOR OF σ_0^2 FOR THE RANDOM MODEL

Under H_0: $\sigma_1^2 = \sigma_2^2 = \cdots = \sigma_N^2 = \sigma_0^2$, V of Eq. 6.B.4 becomes

$$V = \sigma_0^{2N} \tag{6.D.1}$$

and Q of Eq. 6.B.3 is

$$Q = \sigma_0^{2(N-1)}\left(\sum_{j=1}^{N} a_j^2 + \sigma_0^2 \right) \tag{6.D.2}$$

From Eqs. 6.B.5 and 6.B.6, respectively, the elements of the inverse matrix are

$$\sigma^{ii} = \frac{\displaystyle\sum_{j=1}^{N} a_j^2 - a_i^2 + \sigma_0^2}{\sigma_0^2 \left(\displaystyle\sum_{j=1}^{N} a_j^2 + \sigma_0^2 \right)} \tag{6.D.3}$$

and

$$\sigma^{ij} = - \frac{a_i a_j}{\sigma_0^2 \left(\displaystyle\sum_{j=1}^{N} a_j^2 + \sigma_0^2 \right)} \tag{6.D.4}$$

To simplify the notation, let

$$b_0 = \sum_{j=1}^{N} a_j^2 \tag{6.D.5}$$

Then L of Eq. 6.B.7 is rewritten

$$L = C - 0.5n(N-1)\ln \sigma_0^2 - 0.5n \ln\left(b_0 + \sigma_0^2\right) - \frac{0.5(n-1)}{\sigma_0^2\left(b_0 + \sigma_0^2\right)}$$

$$\times \left[\left(b_0 + \sigma_0^2\right)\sum_{j=1}^{N} s_j^2 - \sum_{j=1}^{N} a_j^2 s_j^2 - 2\sum_{i=1}^{N-1}\sum_{j>i}^{N} a_i a_j s_{ij}\right] \tag{6.D.6}$$

To simplify further, let

$$b_1 = \sum_{j=1}^{N} s_j^2 \tag{6.D.7}$$

$$b_2 = \sum_{i=1}^{N-1}\sum_{j>i}^{N} a_i a_j s_{ij} \tag{6.D.8}$$

$$b_3 = \sum_{j=1}^{N} a_j^2 s_j^2 \tag{6.D.9}$$

and then L is rewritten

$$L = C - 0.5n(N-1)\ln \sigma_0^2 - 0.5n \ln\left(b_0 + \sigma_0^2\right)$$

$$- \frac{0.5(n-1)}{\sigma_0^2\left(b_0 + \sigma_0^2\right)}\left[\left(b_0 + \sigma_0^2\right)b_1 - b_3 - 2b_2\right] \tag{6.D.10}$$

To find the MLE of σ_0^2,

$$\frac{\partial L}{\partial \sigma_0^2} = -\frac{0.5n(N-1)}{\sigma_0^2} - \frac{0.5n}{b_0 + \sigma_0^2} - \frac{0.5(n-1)A}{\sigma_0^4\left(b_0 + \sigma_0^2\right)^2} = 0 \tag{6.D.11}$$

where

$$A = -b_1\sigma_0^4 - \left(b_0 b_1 - b_3 - 2b_2\right)\left(b_0 + 2\sigma_0^2\right) \tag{6.D.12}$$

Upon simplifying and combining terms, Eq. 6.D.11 becomes a cubic equation in σ_0^2 of the form:

$$c_3\sigma_0^6 + c_2\sigma_0^4 + c_1\sigma_0^2 + c_0 = 0 \tag{6.D.13}$$

where

$$c_3 = nN \tag{6.D.14}$$

$$c_2 = nb_0(2N - 1) - (n - 1)b_1 \tag{6.D.15}$$

$$c_1 = n(N - 1)b_0^2 - 2(n - 1)(b_0b_1 - b_3 - 2b_2) \tag{6.D.16}$$

$$c_0 = (n - 1)b_0(2b_2 + b_3 - b_0b_1) \tag{6.D.17}$$

DERIVATION OF CONSTRAINED MAXIMUM LIKELIHOOD ESTIMATES FOR RANDOM MODEL

The log likelihood for $\sigma_i^2 = 0$ is easily derived from Eq. 6.B.20:

$$L = C - 0.5n \sum_{j \neq i} \ln \sigma_j^2 - 0.5n \ln b_{0i} - \frac{0.5(n-1)b_{2i}}{b_{0i}} \qquad (6.E.1)$$

Using Eqs. 6.B.16, 6.B.18, 6.B.9 to 6.B.11, and 6.B.13, L is rewritten:

$$L = C - 0.5n \sum_{j \neq i} \ln \sigma_j^2 - 0.5n \ln a_i^2 - 0.5(n-1)$$

$$\times \left[\left(\sum_{j \neq i} a_j^2 \sigma_j^2 + 1 \right) \left(\frac{s_i^2}{a_i^2} \right) + \sum_{j \neq i} \frac{s_j^2}{\sigma_j^2} - 2 \sum_{j \neq i} \frac{a_j s_{ij}}{a_i \sigma_j^2} \right] \qquad (6.E.2)$$

To find the CMLE of σ_k^2 for $k \neq i$,

$$\frac{\partial L}{\partial \sigma_k^2} = -\frac{0.5n}{\sigma_k^2} + 0.5(n-1) \left(\frac{a_k^2 s_i^2}{a_i^2 \sigma_k^4} + \frac{s_k^2}{\sigma_k^4} - \frac{2a_k s_{ik}}{a_i \sigma_k^4} \right) = 0 \qquad (6.E.3)$$

from which

$$\hat{\sigma}_k^2 = \frac{(n-1)\left(a_i^2 s_k^2 + a_k^2 s_i^2 - 2a_i a_k s_{ik} \right)}{na_i^2} \qquad (6.E.4)$$

which is Eq. 6.5.1.

Next, in evaluating L corresponding to the CMLE's of σ_j^2, note that the expression in the square brackets in Eq. 6.E.2 may be written

$$\frac{s_i^2}{a_i^2} + \sum_{j \neq i} \left(\frac{a_j^2 s_i^2}{a_i^2 \sigma_j^2} + \frac{s_j^2}{\sigma_j^2} - \frac{2a_j s_{ij}}{a_i \sigma_j^2} \right) \qquad (6.E.5)$$

which, after factoring out σ_j^2, replacing it by Eq. 6.E.4, and combining terms becomes

$$\frac{s_i^2}{a_i^2} + \sum_{j \neq i} \frac{\left(a_j^2 s_i^2 + a_i^2 s_j^2 - 2a_i a_j s_{ij}\right) na_i^2}{a_i^2 \left(a_i^2 s_j^2 + a_j^2 s_i^2 - 2a_i a_j s_{ij}\right)} = \frac{s_i^2}{a_i^2} + n(N-1) \quad (6.E.6)$$

since there are $(N-1)$ terms in the summation. Thus corresponding to the CMLE's, L of Eq. 6.E.2 is

$$\hat{L} = C - 0.5n \sum_{j \neq i} \ln \hat{\sigma}_j^2 - 0.5n \ln a_i^2$$

$$- \frac{0.5(n-1)s_i^2}{a_i^2} - 0.5n(n-1)(N-1) \quad (6.E.7)$$

Since all terms are independent of $\hat{\sigma}_j^2$ except the summation, it follows that the $\hat{\sigma}_j^2$ that corresponds to the largest likelihood when equated to zero is the one that minimizes

$$\sum_{j \neq i} \ln \hat{\sigma}_j^2$$

This result is the basis for the selection criterion following Eq. 6.5.1.

DERIVATION OF CONSTRAINED MAXIMUM LIKELIHOOD ESTIMATES FOR FIXED MODEL

For the fixed model, μ_k is an unknown constant for each item k. From the model (Eq. 1.8.1), it follows that

$$E(x_{ik}) = \alpha_i + \beta_i \mu_k \tag{6.F.1}$$

Since μ_k is not a random variable and hence x_{ik} and x_{jk} are independently distributed with variances σ_i^2 and σ_j^2, respectively, the variance-covariance matrix (σ_{ij}) is an $N \times N$ diagonal matrix with element σ_i^2 in position (i, i). This leads to the following simple expression for the log likelihood function:

$$L = C - 0.5n \sum_{i=1}^{N} \ln \sigma_i^2 - 0.5 \sum_{i=1}^{N} \frac{\sum_{k=1}^{n} (x_{ik} - \alpha_i - \beta_i \mu_k)^2}{\sigma_i^2} \tag{6.F.2}$$

The constraint is $\sigma_i^2 = 0$. Arbitrarily set $\beta_i = 1$ and $\alpha_i = 0$. This arbitrary action does not affect the estimates of σ_j^2 for $j \neq i$. Now, with $\sigma_i^2 = 0$, $\beta_i = 1$, and $\alpha_i = 0$, it follows immediately that

$$\hat{\mu}_k = x_{ik}; \qquad k = 1, 2, \ldots, n \tag{6.F.3}$$

since method i is accorded infinite weight in estimating μ_k. Upon replacing μ_k in Eq. 6.F.2 by x_{ik}, L is rewritten

$$L = C - 0.5n \sum_{j \neq i} \ln \sigma_j^2 - 0.5 \sum_{j \neq i}^{N} \frac{\sum_{k=1}^{n} (x_{jk} - \alpha_j - \beta_j x_{ik})^2}{\sigma_j^2} \tag{6.F.4}$$

Then,

$$\frac{\partial L}{\partial \alpha_j} = - \frac{\sum_{k=1}^{n} (x_{jk} - \alpha_j - \beta_j x_{ik})}{\sigma_j^2} = 0$$

gives

$$\hat{\alpha}_j = \bar{x}_j - \hat{\beta}_j \bar{x}_i \tag{6.F.5}$$

This is Eq. 6.5.4 in the text.

Also,

$$\frac{\partial L}{\partial \beta_j} = -\frac{\sum_{k=1}^{n} (x_{jk} - \alpha_j - \beta_j x_{ik}) x_{ik}}{\sigma_j^2} = 0$$

Replacing α_j by $\bar{x}_j - \beta_j \bar{x}_i$, this becomes

$$(n-1)(s_{ij} - \beta_j s_i^2) = 0$$

which gives

$$\hat{\beta}_j = \frac{s_{ij}}{s_i^2} \tag{6.F.6}$$

which is Eq. 6.5.3 in the text.

Finally, from Eq. 6.F.4,

$$\frac{\partial L}{\partial \sigma_j^2} = -\frac{0.5n}{\sigma_j^2} + 0.5 \sum_{k=1}^{n} \frac{(x_{jk} - \alpha_j - \beta_j x_{ik})^2}{\sigma_j^4} = 0 \tag{6.F.7}$$

from which

$$\hat{\sigma}_j^2 = \frac{\sum_{k=1}^{n} (x_{jk} - \alpha_j - \beta_j x_{ik})^2}{n} \tag{6.F.8}$$

Replace α_j and β_j by their CMLE's given by Eqs. 6.F.5 and 6.F.6, respectively,

$$\hat{\sigma}_j^2 = \frac{\sum_{k=1}^{n} [(x_{jk} - \bar{x}_j) - \hat{\beta}_j (x_{ik} - \bar{x}_i)]^2}{n}$$

$$= \frac{(n-1)(s_j^2 - 2s_{ij}^2/s_i^2 + s_{ij}^2/s_i^2)}{n}$$

$$= \frac{(s_j^2 - s_{ij}^2/s_i^2)(n-1)}{n} \tag{6.F.9}$$

which is Eq. 6.5.2 in the text, thus completing the derivation.

REFERENCES

1. Grubbs, F. E. "On Estimating Precision of Measuring Instruments and Product Variability." *J. Amer. Stat. Assn.* **43**: 243–264; 1948.

2. Anderson, T. W. and Rubin, H. "Statistical Inference in Factor Analysis." *Proceedings of the Third Berkeley Symposium*; pp. 111–150; 1955.

3. Jaech, J. L. "A Program to Estimate Measurement Errors in Nondestructive Evaluation of Fuel Element Quality." *Technometrics.* **6**: 293–300; 1964.

Chapter 7

COMPARISON OF MOMENT AND MAXIMUM LIKELIHOOD ESTIMATION METHODS

7.1 INTRODUCTION

The Grubbs' (1) approach to parameter estimation is based on equating population moments to sample moments and solving the resulting equations in some sense for the unknown parameters, the estimates of which are then the solutions to these equations. Whenever the number of moment equations is equal to the number of parameters, there is a unique solution to the equations, that is, a unique set of parameter estimates. Whenever the number of equations exceeds the number of parameters, the problem becomes one of how best to "solve" the equations.

Grubbs' estimators, labeled the moment estimates (ME's) in Chapter 6 and henceforth in Chapter 7, are the least squares solutions of the moment equations; in other words, the estimates are chosen to minimize the sum of the squared differences between the population and sample moments. Alternatively, the moment equations may be solved by generalized least squares (2), a solution that weights the equations by taking into account the sampling variances of the sample moments and the sampling covariances between all pairs of sample moments. The resulting estimators, as well as the estimates themselves, are henceforth labeled GLSE's. The estimates derived from applying the maximum likelihood principle continue to be labeled MLE's.

In Chapter 7, ME's, GLSE's, and MLE's are compared. Some of the results from comparing ME's and MLE's given in this chapter were mentioned earlier. These are reviewed and comparisons are given in Section 7.2. A conjecture is then made in Section 7.3 about the relationships among the ME's, GLSE's, and MLE's. Section 7.4 proves this conjecture for a number of specific cases, followed by further demonstration in Section 7.5 for some additional cases.

From the viewpoint of applications, Chapters 1 to 6 contain all the results needed to make inferences about the measurement error parameters. Chapter 7 is intended more for the reader who is interested in reconciling the two estimation procedures. It contains no results that would be used in an application, but it does show how the moment equations may be solved to provide parameter estimates in essential agreement with the MLE's. This correspondence between two quite different approaches to estimation is certainly of interest, if not of practical importance. The reader who wishes to pursue this area in more mathematical depth is referred to a paper by Browne (3) that discusses the relationships between MLE's and GLSE's.

7.2   COMPARISONS BETWEEN MOMENT AND MAXIMUM LIKELIHOOD ESTIMATORS

Table 7.1 reviews results from earlier chapters showing the relationship between the ME's and MLE's in specific applications. In Table 7.1, N is the number of measurement methods and n the number of items measured. Throughout, the random model is assumed for $N = 2$, while the model may be either random or fixed for $N \geq 3$, since attention is restricted to estimation based on paired differences. In the examples of Chapter 6 that deal with the nonconstant bias model, we noted that there may be appreciable differences between the ME's and MLE's. Because of the mathematical complexity of the nonconstant bias model, this ME-MLE comparison is carried no further here. However, conjectures are made about the nonconstant bias model. ME's and MLE's are compared further in the following sections.

7.2.1 $N = 2$; $\sigma_1^2 = 0$

In Section 3.5.1, the MLE's of σ_2^2 and σ_μ^2 for this model were given as

$$\hat{\sigma}_2^2 = \frac{(n-1)\left(s_1^2 - 2s_{12} + s_2^2\right)}{n} \tag{7.2.1}$$

and

$$\hat{\sigma}_\mu^2 = \frac{(n-1)s_1^2}{n} \tag{7.2.2}$$

Table 7.1   Comparisons of ME's and MLE's for Applications in Chapters 3 to 5

Section	Model Description	Comparison
3.3.1	$N = 2$; $\sigma_1^2 \neq \sigma_2^2$	ME's of σ_1^2, σ_2^2, and σ_μ^2 are $n/(n-1)$ times MLE's
4.3	$N = 3$; $\sigma_1^2 \neq \sigma_2^2 \neq \sigma_3^2$	ME of σ_i^2 is $n/(n-1)$ times MLE for all i
5.3	$N \geq 4$; $\sigma_1^2 \neq \sigma_2^2 \neq \cdots \neq \sigma_N^2$	ME of σ_i^2 may differ appreciably from MLE for all i

The ME's are found by solving Eq. 3.2.2 by least squares, having set $\sigma_1^2 = 0$. The equations reduce to

$$s_1^2 = \sigma_\mu^2$$

$$s_2^2 = \sigma_2^2 + \sigma_\mu^2 \qquad (7.2.3)$$

$$s_{12} = \sigma_\mu^2$$

Appendix 7.A shows that the ME's of σ_2^2 and σ_μ^2 are

$$\tilde{\sigma}_2^2 = 0.5\left(2s_2^2 - s_1^2 - s_{12}\right) \qquad (7.2.4)$$

$$\tilde{\sigma}_\mu^2 = 0.5\left(s_1^2 + s_{12}\right) \qquad (7.2.5)$$

Note that the estimating Eqs. 7.2.4 and 7.2.5 are quite different from Eqs. 7.2.1 and 7.2.2, respectively; that is, the ME's and the MLE's are not related within the factor $(n - 1)/n$ in this instance.

7.2.2 $N \geq 3$; $\sigma_i^2 = 0$

In Section 4.5 for $N = 3$ and Section 5.5 for $N \geq 4$, the MLE of σ_j^2 for $j \neq i$ is given as

$$\hat{\sigma}_j^2 = \frac{(n - 1)V_{ij}}{n}; \qquad j \neq i \qquad (7.2.6)$$

The ME of σ_j^2 for $j \neq i$ is derived in Appendix 7.B to be

$$\tilde{\sigma}_j^2 = \frac{2(N - 2)\sum\limits_{\substack{k=1 \\ \neq j}}^{N} V_{jk} - \sum\limits_{\substack{l \neq i, j}}^{N} \sum\limits_{\substack{k=1 \\ \neq l}}^{N} V_{lk}}{(N - 2)(2N - 3)} \qquad (7.2.7)$$

For example, if $N = 4$ and $i = 3$; that is, $\sigma_3^2 = 0$, then the ME of σ_2^2, say, is

$$\tilde{\sigma}_2^2 = \frac{4(V_{12} + V_{23} + V_{24}) - \left[(V_{12} + V_{13} + V_{14}) + (V_{14} + V_{24} + V_{34})\right]}{10}$$

Again, for this case the MLE given by Eq. 7.2.6 differs appreciably from the ME given by Eq. 7.2.7.

EXAMPLE 7.A

Example 5.A dealt with the measurement of 43 fuel pellet densities by each of 6 measurement methods. For these data, find the estimates of $\sigma_2^2, \sigma_3^2, \ldots, \sigma_6^2$ if $\sigma_1^2 = 0$. The V_{ij} values are given in the cited example. The variances are all expressed as $(\%TD)^2$.

From Eq. 7.2.6, the MLE's are

$$\hat{\sigma}_2^2 = \frac{42(0.015412)}{43} = 0.015054$$

$$\hat{\sigma}_3^2 = 0.029668 \qquad \hat{\sigma}_5^2 = 0.113681$$

$$\hat{\sigma}_4^2 = 0.071415 \qquad \hat{\sigma}_6^2 = 0.043467$$

The ME's are given by Eq. 7.2.7. Sums used in the calculations are as follows:

$$\sum_{\substack{k=1 \\ \neq 2}}^{6} V_{2k} = 0.317311 \qquad \sum_{\substack{k=1 \\ \neq 5}}^{6} V_{5k} = 0.694521$$

$$\sum_{\substack{k=1 \\ \neq 3}}^{6} V_{3k} = 0.342490 \qquad \sum_{\substack{k=1 \\ \neq 6}}^{6} V_{6k} = 0.432561$$

$$\sum_{\substack{k=1 \\ \neq 4}}^{6} V_{4k} = 0.528166$$

Then,

$$\tilde{\sigma}_2^2 = \frac{8(0.317311) - (0.342490 + 0.528166 + 0.694521 + 0.432561)}{36}$$

$$= 0.015021$$

$$\tilde{\sigma}_3^2 = 0.021316 \qquad \tilde{\sigma}_5^2 = 0.109323$$

$$\tilde{\sigma}_4^2 = 0.067735 \qquad \tilde{\sigma}_6^2 = 0.043833 \qquad \blacksquare$$

7.2.3 $N \geq 3$; $\sigma_i^2 = \sigma_0^2$ for All i

In Section 4.4.2 for $N = 3$ and Section 5.4.2 for $N \geq 4$, it is shown that the MLE of σ_0^2 is

$$\hat{\sigma}_0^2 = \frac{(n-1) \sum_{i=1}^{N-1} \sum_{j>i} V_{ij}}{nN(N-1)} \qquad (7.2.8)$$

The ME of σ_0^2 is derived by noting that the least squares estimate (LSE) is found by minimizing

$$\sum_{i=1}^{N-1} \sum_{j>i} \left(V_{ij} - 2\sigma_0^2 \right)^2$$

which leads immediately to the ME:

$$\tilde{\sigma}_0^2 = \frac{\displaystyle\sum_{i=1}^{N-1} \sum_{j>i} V_{ij}}{N(N-1)} \tag{7.2.9}$$

upon noting that there are $N(N-1)/2$ terms in the double summation.

Note that for this case, the ME and MLE are identical within the factor $(n-1)/n$.

7.2.4 $N = 3$; $\sigma_1^2 = \sigma_2^2 = \sigma_0^2$; $\sigma_3^2 \neq \sigma_0^2$

For this case, there are two parameters to estimate and three moment equations. The MLE's of σ_0^2 and σ_3^2 are derived in Appendix 7.C. They are, respectively,

$$\hat{\sigma}_0^2 = \frac{(n-1)V_{12}}{2n} \tag{7.2.10}$$

$$\hat{\sigma}_3^2 = \frac{(n-1)(V_{13} + V_{23} - V_{12})}{2n} \tag{7.2.11}$$

The ME's of σ_0^2 and σ_3^2 are the solutions of the matrix equation

$$\begin{pmatrix} 6 & 2 \\ 2 & 2 \end{pmatrix} \begin{pmatrix} \sigma_0^2 \\ \sigma_3^2 \end{pmatrix} = \begin{pmatrix} 2V_{12} + V_{13} + V_{23} \\ V_{13} + V_{23} \end{pmatrix} \tag{7.2.12}$$

which are

$$\tilde{\sigma}_0^2 = \frac{V_{12}}{2} \tag{7.2.13}$$

$$\tilde{\sigma}_3^2 = \frac{V_{13} + V_{23} - V_{12}}{2} \tag{7.2.14}$$

In comparing Eqs. 7.2.13 and 7.2.14 with Eqs. 7.2.10 and 7.2.11, respectively, note that for this example, the ME's and the MLE's are again related within the $(n-1)/n$ factor.

This completes the comparisons between the ME's and the MLE's in this section; one additional example is given later in Section 7.5.3. The next section notes the characteristics of those cases in which the ME's agree with the MLE's within the $(n - 1)/n$ factor. A conjecture is made about the conditions under which these estimators do agree within the $(n - 1)/n$ factor, followed by a further conjecture about how the moment equations may be solved, in general, to produce estimators in agreement with the MLE's.

7.3   CONJECTURES ABOUT METHODS OF PARAMETER ESTIMATION

Table 7.2 summarizes the results of Section 7.2 by identifying characteristics of the cases that may be relevant to determining whether the ME's and the MLE's are in agreement. [Henceforth, we label the two estimators as being "in agreement" if they differ only by the $(n - 1)/n$ factor]. For $N = 2$, the number of parameters to estimate is 3 (σ_1^2, σ_2^2, and σ_μ^2), while for $N \geq 3$, there are N parameters, counting only the estimators of σ_i^2.

In determining when there is agreement between the ME's and MLE's, note first that whenever the number of equations equals the number of parameters (cases 1 and 2), there is agreement. However, if the number of equations exceeds the number of parameters, there may be agreement (cases 6 and 7), or there may not (cases 3, 4, and 5).

To pursue this characterization further, note first that for case 6, all of the moment equations are of the form

$$V_{ij} = 2\sigma_0^2 \tag{7.3.1}$$

and hence, since V_{ij} has an expected value that is independent of i and j, the sampling variance of V_{ij} is the same for all pairs i and j; all $N(N - 1)/2$ moment equations are assigned equal weight.

Table 7.2   Characteristics of Cases in Which ME's and MLE's Were Compared in Section 7.2

Case Identification	Number of Moment Equations	Number of Parameters	Agreement Between ME or MLE?
1) Table 7.1; 3.3.1	3	3	Yes
2) Table 7.1; 4.3	3	3	Yes
3) Table 7.1; 5.3	$N(N - 1)/2$	N	No
4) 7.2.1	3	2	No
5) 7.2.2	$N(N - 1)/2$	$N - 1$	No
6) 7.2.3	$N(N - 1)/2$	1	Yes
7) 7.2.4	3	2	Yes

Turning to case 7, note that the three moment equations are

$$V_{12} = 2\sigma_0^2$$
$$V_{13} = \sigma_0^2 + \sigma_3^2 \tag{7.3.2}$$
$$V_{23} = \sigma_0^2 + \sigma_3^2$$

Again, noting that the last two equations have equal weight, they may be combined to form the equation

$$V_{13} + V_{23} = 2(\sigma_0^2 + \sigma_3^2) \tag{7.3.3}$$

and there are now two equations in two unknowns, the same number of equations as there are unknown parameters.

For cases 3 to 5, the number of equations cannot be reduced to the number of unknowns by adding equations having the same sampling variance. For case 3, all V_{ij}'s have unequal sampling variances; no equations may be added. For case 4, the three moment equations are

$$s_1^2 = \sigma_\mu^2$$
$$s_2^2 = \sigma_\mu^2 + \sigma_2^2 \tag{7.3.4}$$
$$s_{12} = \sigma_\mu^2$$

and, although both s_1^2 and s_{12} have the same expected value, σ_μ^2, they do not have the same sampling variance and cannot simply be added to reduce the number of equations to two. For case 5, again all V_{ij}'s have unequal sampling variances.

Examination of these cases leads to the first conjecture.

Conjecture 1.   The ME's and the MLE's are in agreement if (1) the number of moment equations equals the number of parameters to be estimated, or if (2) having added the equations wherein the sample moments have the same expected values and sampling variances, the number of remaining equations equals the number of parameters. [Note, for clarification, that in finding the ME's one does not actually add the equations as point (2) might imply; the original equations are retained].

Since the key to combining equations is that they have the same sampling variance, one naturally asks how the moment equations may be solved to produce estimators in agreement with the MLE's when the conditions of Conjecture 1 are not met. A logical approach is to solve the equations by generalized least squares. Calling the resulting estimators, and also the estimates, GLSE's, a second conjecture is developed.

Conjecture 2.   If the moment equations are solved by generalized least squares, then the GLSE's are in agreement with the MLE's.

This conjecture also implies that if the conditions of Conjecture 1 are met, then the ME's, GLSE's, and MLE's are all in agreement.

Conjecture 2 is proved for a number of specific cases in Section 7.4 and demonstrated by examples for a number of other cases in Section 7.5.

7.4   PROOFS OF CONJECTURE 2 FOR SPECIFIC CASES

First, the GSLE's are given in matrix notation. Let the number of moment equations be m and the number of parameters be $p \leq m$. The moment equations are written in the form

$$T'(\sigma^2)' = Y' \tag{7.4.1}$$

where T' is the $m \times p$ matrix of coefficients, $(\sigma^2)'$ is the p-dimensional column matrix of unknown parameters, and Y' is the m-dimensional column matrix of sample moments. Let V be the $m \times m$ variance-covariance matrix of sampling variances and covariances. Then, Aitken (2) gives the generalized least squares solution to Eq. 7.4.1 as

$$(\dot{\sigma}^2) = YV^{-1}T'(TV^{-1}T')^{-1} \tag{7.4.2}$$

Conjecture 2 is now proved for a number of specific cases upon application of Eq. 7.4.2 with $(\dot{\sigma}^2)$ identified as the p-dimensional row matrix of GLSE's of the parameters.

7.4.1 $N = 2$; $\sigma_1^2 = 0$

This is the case of Section 7.2.1, for which the ME's and MLE's were not in agreement. For this case, $m = 3$, $p = 2$, and the matrices in Eq. 7.4.1 are, from Eq. 7.3.4,

$$T' = \begin{pmatrix} 0 & 1 \\ 1 & 1 \\ 0 & 1 \end{pmatrix} \tag{7.4.3}$$

$$(\sigma^2)' = \begin{pmatrix} \sigma_\mu^2 \\ \sigma_2^2 \end{pmatrix} \tag{7.4.4}$$

$$Y' = \begin{pmatrix} s_1^2 \\ s_2^2 \\ s_{12} \end{pmatrix} \tag{7.4.5}$$

The elements of the V matrix are based on the following results for the assumed random model:

$$\text{var } s_i^2 = \frac{2\left(\sigma_\mu^2 + \sigma_i^2\right)^2}{n-1}; \qquad i = 1,2 \tag{7.4.6}$$

$$\text{var } s_{12} = \frac{2\sigma_\mu^4 + \sigma_\mu^2\sigma_1^2 + \sigma_\mu^2\sigma_2^2 + \sigma_1^2\sigma_2^2}{(n-1)} \tag{7.4.7}$$

$$\text{cov}\left(s_1^2, s_2^2\right) = \frac{2\sigma_\mu^4}{(n-1)} \tag{7.4.8}$$

$$\text{cov}\left(s_i^2, s_{12}\right) = 2\sigma_\mu^2\left(\sigma_\mu^2 + \sigma_i^2\right); \qquad i = 1,2 \tag{7.4.9}$$

For $\sigma_1^2 = 0$, the V matrix is

$$V = \frac{1}{n-1} \begin{pmatrix} 2\sigma_\mu^4 & 2\sigma_\mu^4 & 2\sigma_\mu^4 \\ & 2\left(\sigma_\mu^2 + \sigma_2^2\right)^2 & 2\sigma_\mu^2\left(\sigma_\mu^2 + \sigma_2^2\right) \\ \text{symmetric matrix} & & \sigma_\mu^2\left(2\sigma_\mu^2 + \sigma_2^2\right) \end{pmatrix} \tag{7.4.10}$$

In applying Eq. 7.4.2, the matrix operations are straightforward. As intermediate results we get

$$\left(TV^{-1}T'\right)^{-1} = \frac{2}{n-1} \begin{pmatrix} \sigma_\mu^4 & 0 \\ 0 & \sigma_2^4 \end{pmatrix} \tag{7.4.11}$$

and

$$YV^{-1}T' = 0.5(n-1)\left(\frac{s_1^2}{\sigma_\mu^4} \quad \frac{\left(s_1^2 + s_2^2 - 2s_{12}\right)}{\sigma_2^4} \right) \tag{7.4.12}$$

It follows immediately from Eqs. 7.4.2, 7.4.11, and 7.4.12 that the GSLE's of σ_μ^2 and σ_2^2 are

$$\dot{\sigma}_\mu^2 = s_1^2$$

$$\dot{\sigma}_2^2 = s_1^2 + s_2^2 - 2s_{12} \tag{7.4.13}$$

From Eqs. 7.2.1 and 7.2.2, it is noted that these are the MLE's within the $(n-1)/n$ factor. Conjecture 2 is proven for this special case.

7.4.2 $N = 3$; $\sigma_1^2 = 0$

This is a special case of the Section 7.2.2 case with $N = 3$ rather then $N \geq 3$. The parameter σ_1^2 is set equal to zero without loss of generality. For this case, $m = 3$, $p = 2$, and the matrices in Eq. 7.4.1 are

$$T' = \begin{pmatrix} 1 & 0 \\ 0 & 1 \\ 1 & 1 \end{pmatrix} \tag{7.4.14}$$

$$(\sigma^2)' = \begin{pmatrix} \sigma_2^2 \\ \sigma_3^2 \end{pmatrix} \tag{7.4.15}$$

$$Y' = \begin{pmatrix} V_{12} \\ V_{13} \\ V_{23} \end{pmatrix} \tag{7.4.16}$$

The elements of V are based on the following results:

$$\text{var } V_{ij} = \frac{2\left(\sigma_i^2 + \sigma_j^2\right)^2}{n-1} \tag{7.4.17}$$

$$\text{cov}\left(V_{ij}, V_{ik}\right) = \frac{2\sigma_i^4}{n-1} \tag{7.4.18}$$

For $\sigma_1^2 = 0$, the V matrix is

$$V = \frac{2}{n-1} \begin{pmatrix} \sigma_2^4 & 0 & \sigma_2^4 \\ & \sigma_3^4 & \sigma_3^4 \\ \text{symmetric matrix} & & \left(\sigma_2^2 + \sigma_3^2\right)^2 \end{pmatrix} \tag{7.4.19}$$

Again, the matrix operations are straightforward. The intermediate results are

$$(TV^{-1}T')^{-1} = \frac{2}{n-1} \begin{pmatrix} \sigma_2^4 & 0 \\ 0 & \sigma_3^4 \end{pmatrix} \tag{7.4.20}$$

and

$$YV^{-1}T' = 0.5(n-1)\left(\frac{V_{12}}{\sigma_2^4} \quad \frac{V_{13}}{\sigma_3^4}\right) \tag{7.4.21}$$

From Eqs. 7.4.2, 7.4.20, and 7.4.21, the GSLE's of σ_2^2 and σ_3^2 are

$$\dot{\sigma}_2^2 = V_{12}$$
$$\dot{\sigma}_3^2 = V_{13} \tag{7.4.22}$$

From Eq. 7.2.6 for $i = 1$ and $j = 2$ and 3, respectively, it is noted that these are the MLE's within the $(n-1)/n$ factor. Conjecture 2 is proven for this special case.

7.4.3 $N = 3$; $\sigma_1^2 = \sigma_2^2 = \sigma_0^2$; $\sigma_3^2 \neq \sigma_0^2$

This is the case of Section 7.2.4, which is of special interest because the ME's and MLE's were in agreement. For this case, $m = 3$ and $p = 2$. The matrices in Eqs. 7.4.1 are, from Eq. 7.3.2,

$$T' = \begin{pmatrix} 2 & 0 \\ 1 & 1 \\ 1 & 1 \end{pmatrix} \tag{7.4.23}$$

$$(\sigma^2)' = \begin{pmatrix} \sigma_0^2 \\ \sigma_3^2 \end{pmatrix} \tag{7.4.24}$$

$$Y' = \begin{pmatrix} V_{12} \\ V_{13} \\ V_{23} \end{pmatrix} \tag{7.4.25}$$

From Eqs. 7.4.17 and 7.4.18,

$$V = \frac{2}{n-1}\begin{pmatrix} 4\sigma_0^4 & \sigma_0^4 & \sigma_0^4 \\ & \left(\sigma_0^2 + \sigma_3^2\right)^2 & \sigma_3^4 \\ \text{symmetric matrix} & & \left(\sigma_0^2 + \sigma_3^2\right)^2 \end{pmatrix} \tag{7.4.26}$$

In Appendix 7.D, it is shown that

$$V^{-1}T'(TV^{-1}T')^{-1} = \begin{pmatrix} 0.5 & -0.5 \\ 0 & 0.5 \\ 0 & 0.5 \end{pmatrix} \tag{7.4.27}$$

and, hence, from Eq. 7.4.2,

$$\dot{\sigma}_0^2 = \frac{V_{12}}{2}$$

$$\dot{\sigma}_3^2 = \frac{-V_{12} + V_{13} + V_{23}}{2} \qquad (7.4.28)$$

From Eqs. 7.2.13 and 7.2.14, it is noted that the GLSE's are identical to the ME's and thus are in agreement with the MLE's. Conjecture 2 is proven for this special case.

Having proven Conjecture 2 for three special cases, its truth will now be demonstrated for a number of other examples. Because of the algebraic complexity in inverting matrices of dimensions greater than 3, three special cases for which Conjecture 2 was proven all had $m = 3$. In the numerical examples to follow, m exceeds 3.

7.5 DEMONSTRATION OF TRUTH OF CONJECTURE 2 BY EXAMPLES

7.5.1 $N = 4$; $\sigma_1^2 \neq \sigma_2^2 \neq \sigma_3^2 \neq \sigma_4^2$

Example 5.C is revisited. In this example, $N = 4$ and $n = 8$. The MLE's were found to be

$$\hat{\sigma}_1^2 = 4.6655 \qquad \hat{\sigma}_3^2 = 0.4486$$

$$\hat{\sigma}_2^2 = 1.0505 \qquad \hat{\sigma}_4^2 = 1.2059$$

The ME's were found to be

$$\tilde{\sigma}_1^2 = 5.3838 \qquad \tilde{\sigma}_3^2 = 0.5142$$

$$\tilde{\sigma}_2^2 = 0.0606 \qquad \tilde{\sigma}_4^2 = 2.5224$$

Note that these estimates differ appreciably from the MLE's; note especially that for $i = 2$, the ratio $\hat{\sigma}_2^2/\tilde{\sigma}_2^2$ exceeds 17.

To find the GLSE's using Eq. 7.4.2, the following matrices are defined:

$$T' = \begin{pmatrix} 1 & 1 & 0 & 0 \\ 1 & 0 & 1 & 0 \\ 1 & 0 & 0 & 1 \\ 0 & 1 & 1 & 0 \\ 0 & 1 & 0 & 1 \\ 0 & 0 & 1 & 1 \end{pmatrix} \tag{7.5.1}$$

$$(\sigma^2) = \begin{pmatrix} \sigma_1^2 & \sigma_2^2 & \sigma_3^2 & \sigma_4^2 \end{pmatrix} \tag{7.5.2}$$

$$Y = \begin{pmatrix} V_{12} & V_{13} & V_{14} & V_{23} & V_{24} & V_{34} \end{pmatrix} \tag{7.5.3}$$

$V = \dfrac{2}{7}$

$$\times \begin{pmatrix} (\sigma_1^2 + \sigma_2^2)^2 & \sigma_1^4 & \sigma_1^4 & \sigma_2^4 & \sigma_2^4 & 0 \\ & (\sigma_1^2 + \sigma_3^2)^2 & \sigma_1^4 & \sigma_3^4 & 0 & \sigma_3^4 \\ & & (\sigma_1^2 + \sigma_4^2)^2 & 0 & \sigma_4^4 & \sigma_4^4 \\ & \text{symmetric matrix} & & (\sigma_2^2 + \sigma_3^2)^2 & \sigma_2^4 & \sigma_3^4 \\ & & & & (\sigma_2^2 + \sigma_4^2)^2 & \sigma_4^4 \\ & & & & & (\sigma_3^2 + \sigma_4^2)^2 \end{pmatrix}$$

$$\tag{7.5.4}$$

The difficulties in inverting V algebraically to apply Eq. 7.4.2 are apparent. Note that in the examples of Section 7.4, the elements of V, which are the true but unknown parameter values, cancelled out in the final steps; in other words, it was not necessary to know the elements of V in order to calculate the GLSE's. The same is not true here; that is, the right-hand side (RHS) of Eq. 7.4.2 depends on the elements of V. However, a unique solution for $(\dot{\sigma})^2$ can be arrived at iteratively, and this unique solution becomes the GLSE's.

Initial values are assigned σ_i^2 for $i = 1, 2, 3, 4$ to obtain an initial set of estimates by applying Eq. 7.4.2. These estimates are then inputs in the second round to provide the elements of V and a second set of estimates. The iterative process stops when the input σ_i^2 values equal the output σ_i^2 for all i within specified criteria. Experience shows that convergence is rapid.

Following this procedure with the assistance of a computer, the resulting GLSE's for this example were

$$\dot{\sigma}_1^2 = 5.3318 \qquad \dot{\sigma}_3^2 = 0.5128$$

$$\dot{\sigma}_2^2 = 1.2001 \qquad \dot{\sigma}_4^2 = 1.3782$$

Note that $\dot{\sigma}_i^2$ is $n/(n-1)$ or $8/7$ times $\hat{\sigma}_i^2$ for all i, thus demonstrating the truth of Conjecture 2 for this example.

Table 7.3   ME's and MLE's for Example 5.A

i	ME $(\tilde{\sigma}_i^2)$	MLE $(\hat{\sigma}_i^2)$
1	0.005077	0.006151
2	0.014457	0.008628
3	0.020752	0.024503
4	0.067171	0.072838
5	0.108759	0.113986
6	0.043269	0.031918

7.5.2 $N = 6$, $\sigma_1^2 \neq \sigma_2^2 \neq \cdots \neq \sigma_6^2$

Example 5.A is revisited. In this example, $N = 6$ and $n = 43$. The ME's and MLE's are as given in Table 7.3. Again, note the differences in the estimates.

In finding the GLSE's, the matrices T', (σ^2), Y, and V are similar to those given by Eqs. 7.5.1 to 7.5.4, except that now T' is 15×6, (σ^2) is 1×6, Y is 1×15, and V is 15×15 with elements suggested by Eq. 7.5.4 for the $N = 4$ case.

By again performing the iterative estimation, the GSLE's are found to be

$$\dot{\sigma}_1^2 = 0.006297 \qquad \dot{\sigma}_4^2 = 0.074572$$

$$\dot{\sigma}_2^2 = 0.008833 \qquad \dot{\sigma}_5^2 = 0.116700$$

$$\dot{\sigma}_3^2 = 0.025086 \qquad \dot{\sigma}_6^2 = 0.032678$$

Again, note that $\dot{\sigma}_i^2$ is $n/(n-1)$ or $43/42$ times $\hat{\sigma}_i^2$ for all i, demonstrating the truth of Conjecture 2 for this example.

7.5.3 $N = 6$; $\sigma_1^2 = \sigma_2^2 = \sigma_3^2 = \delta_1^2$; $\sigma_4^2 = \sigma_5^2 = \delta_2^2$; $\sigma_6^2 = \delta_3^2$

This example was considered in the hypothesis test identified as H_{03} by Jaech (4). The ME's are found from the following 15 equations in 3 unknowns:

$$2\delta_1^2 = 1.32$$

$$2\delta_1^2 = 2.57$$

$$\delta_1^2 + \delta_2^2 = 9.25$$

$$\vdots \qquad \vdots \qquad\qquad (7.5.5)$$

$$2\delta_2^2 = 10.44$$

$$\delta_2^2 + \delta_3^2 = 11.96$$

$$\delta_2^2 + \delta_3^2 = 10.31$$

Note that there are three equations for which $V_{ij} = 2\delta_1^2$; six for which it equals $(\delta_1^2 + \delta_2^2)$; three for which it equals $(\delta_1^2 + \delta_3^2)$; one for which it equals $2\delta_2^2$; and two for which it equals $(\delta_2^2 + \delta_3^2)$. After those sets of equations are combined, there remain five equations in three unknowns. Therefore, by Conjecture 1, we would not expect the ME's and the MLE's to be in agreement.

The ME's are the solutions of the following equations expressed in matrix form:

$$
\begin{pmatrix} 21 & 6 & 3 \\ 6 & 12 & 2 \\ 3 & 2 & 5 \end{pmatrix}
\begin{pmatrix} \delta_1^2 \\ \delta_2^2 \\ \delta_3^2 \end{pmatrix}
=
\begin{pmatrix} 95.62 \\ 111.40 \\ 36.90 \end{pmatrix}
\tag{7.5.6}
$$

$$
\begin{pmatrix} \delta_1^2 \\ \delta_2^2 \\ \delta_3^2 \end{pmatrix}
= \frac{1}{960}
\begin{pmatrix} 56 & -24 & -24 \\ -24 & 96 & -24 \\ -24 & -24 & 216 \end{pmatrix}
\begin{pmatrix} 95.62 \\ 111.40 \\ 36.90 \end{pmatrix}
\tag{7.5.7}
$$

$$
\tilde{\delta}_1^2 = 1.8703
$$

$$
\tilde{\delta}_2^2 = 7.8270
$$

$$
\tilde{\delta}_3^2 = 3.1270
$$

A computer program was used to find the GLSE's. The matrix T' is 15×3, with elements suggested by the left-hand side (LHS) of Eq. 7.5.5, and is of the following form:

$$
T' =
\begin{pmatrix}
2 & 0 & 0 \\
2 & 0 & 0 \\
1 & 1 & 0 \\
\vdots & \vdots & \vdots \\
0 & 2 & 0 \\
0 & 1 & 1 \\
0 & 1 & 1
\end{pmatrix}
\tag{7.5.8}
$$

The (σ^2) matrix is

$$
(\sigma^2) = \begin{pmatrix} \delta_1^2 & \delta_2^2 & \delta_3^2 \end{pmatrix}
\tag{7.5.9}
$$

and Y is

$$
Y = \begin{pmatrix} V_{12} & V_{13} & V_{14} & \cdots & V_{45} & V_{46} & V_{56} \end{pmatrix}
\tag{7.5.10}
$$

The V matrix is 15×15 with elements suggested by Eq. 7.5.4, but with σ_1^2, σ_2^2, and σ_3^2 replaced by δ_1^2; σ_4^2 and σ_5^2 by δ_2^2; and σ_6^2 by δ_3^2.

$$
V = \frac{2}{42}
\begin{pmatrix}
4\delta_1^4 & \delta_1^4 & \delta_1^4 & \cdots & 0 & 0 & 0 \\
 & 4\delta_1^4 & \delta_1^4 & \cdots & 0 & 0 & 0 \\
 & & \left(\delta_1^2 + \delta_2^2\right)^2 & \cdots & \delta_2^4 & \delta_2^4 & 0 \\
 & & & \cdots & & & \\
 & & & \cdots & & & \\
 & \text{symmetric matrix} & & & 4\delta_2^4 & \delta_2^4 & \delta_2^4 \\
 & & & & & \left(\delta_2^2 + \delta_3^2\right)^2 & \delta_3^4 \\
 & & & & & & \left(\delta_2^2 + \delta_3^2\right)^2
\end{pmatrix}
$$

$$(7.5.11)$$

The GSLE's are found to be

$$\hat{\delta}_1^2 = 1.1065$$

$$\hat{\delta}_2^2 = 9.7051$$

$$\hat{\delta}_3^2 = 3.6049$$

and it is noted that these are quite different from the ME's.

Finally, to find the MLE's, the likelihood function Eq. 5.B.1 is rewritten to correspond to the parameter reidentifications and the V_{ij} values for this example:

$$
L = C - 21.5 \ln Q - 21\left(\frac{V}{Q}\right)\left(\frac{6.37}{\delta_1^4} + \frac{68.25}{\delta_1^2 \delta_2^2} + \frac{14.63}{\delta_1^2 \delta_3^2} + \frac{10.44}{\delta_2^4} + \frac{22.27}{\delta_2^2 \delta_3^2}\right)
$$

$$(7.5.12)$$

where

$$V = \delta_1^6 \delta_2^4 \delta_3^2 \tag{7.5.13}$$

and

$$Q = V\left(\frac{3}{\delta_1^2} + \frac{2}{\delta_2^2} + \frac{1}{\delta_3^2}\right) \qquad (7.5.14)$$

The function L may be rewritten

$$L = C - 64.5 \ln \delta_1^2 - 43 \ln \delta_2^2 - 21.5 \ln \delta_3^2 - 21.5 \ln A - \frac{21B}{A} \qquad (7.5.15)$$

where

$$A = \left(\frac{3}{\delta_1^2} + \frac{2}{\delta_2^2} + \frac{1}{\delta_3^2}\right) \qquad (7.5.16)$$

$$B = \left(\frac{6.37}{\delta_1^4} + \frac{68.25}{\delta_1^2\delta_2^2} + \cdots + \frac{22.27}{\delta_2^2\delta_3^2}\right) \qquad (7.5.17)$$

The partial derivatives are

$$\frac{\partial L}{\partial \delta_1^2} = -\frac{64.5}{\delta_1^2} + \frac{64.5}{A\delta_1^4}$$

$$-\frac{21\left[A\left(-12.74/\delta_1^6 - 68.25/\delta_1^4\delta_2^2 - 14.63/\delta_1^4\delta_3^2\right) + 3B/\delta_1^4\right]}{A^2}$$

$$(7.5.18)$$

$$\frac{\partial L}{\partial \delta_2^2} = -\frac{43}{\delta_2^2} + \frac{43}{A\delta_2^4}$$

$$-\frac{21\left[A\left(-68.25/\delta_1^2\delta_2^4 - 20.88/\delta_2^6 - 22.27/\delta_2^4\delta_3^2\right) + 2B/\delta_2^4\right]}{A^2}$$

$$(7.5.19)$$

$$\frac{\partial L}{\partial \delta_3^2} = -\frac{21.5}{\delta_3^2} + \frac{21.5}{A\delta_3^4} - \frac{21\left[A\left(-14.63/\delta_1^2\delta_3^4 - 22.27/\delta_2^2\delta_3^4\right) + B/\delta_3^4\right]}{A^2}$$

$$(7.5.20)$$

Conjecture 2 tells us that the MLE's would be

$$\hat{\delta}_1^2 = \frac{42(1.1065)}{43} = 1.0808$$

$$\hat{\delta}_2^2 = \frac{42(9.7051)}{43} = 9.4794$$

$$\hat{\delta}_3^2 = \frac{42(3.6049)}{43} = 3.5211$$

When these values are inserted in Eqs. 7.5.18 to 7.5.20, the expressions all equal zero. Thus, these values are indeed MLE's, since $\partial L/\partial \delta_i^2 = 0$ for $i = 1, 2, 3$ at these values of the parameters, and the validity of Conjecture 2 is demonstrated for this example.

DERIVATION OF MOMENT ESTIMATORS FOR $N = 2$; $\sigma_1^2 = 0$: EQUATIONS 7.2.4 AND 7.2.5

The quantity to be minimized is

$$Q = \left(s_1^2 - \sigma_\mu^2\right)^2 + \left(s_2^2 - \sigma_2^2 - \sigma_\mu^2\right)^2 + \left(s_{12} - \sigma_\mu^2\right)^2 \qquad (7.A.1)$$

In matrix form, the least squares equations are written

$$\begin{pmatrix} 1 & 1 \\ 1 & 3 \end{pmatrix} \begin{pmatrix} \sigma_2^2 \\ \sigma_\mu^2 \end{pmatrix} = \begin{pmatrix} s_2^2 \\ s_1^2 + s_2^2 + s_{12} \end{pmatrix} \qquad (7.A.2)$$

which yields the solution

$$\begin{pmatrix} \sigma_2^2 \\ \sigma_\mu^2 \end{pmatrix} = \frac{1}{2} \begin{pmatrix} 3 & -1 \\ -1 & 1 \end{pmatrix} \begin{pmatrix} s_2^2 \\ s_1^2 + s_2^2 + s_{12} \end{pmatrix} \qquad (7.A.3)$$

or

$$\tilde{\sigma}_2^2 = 0.5\left(2s_2^2 - s_1^2 - s_{12}\right)$$

$$\tilde{\sigma}_\mu^2 = 0.5\left(s_1^2 + s_{12}\right) \qquad (7.A.4)$$

These are Eqs. 7.2.4 and 7.2.5 in the text.

DERIVATION OF MOMENT ESTIMATORS FOR $N \geq 3$; $\sigma_1^2 = 0$: EQUATION 7.2.7

The least squares equations in matrix notation are of the form

$$B(\sigma^2) = Y \tag{7.B.1}$$

where B is an $(N - 1)$ square matrix with diagonal element $(N - 1)$ and off-diagonal element equal to unity; (σ^2) is an $(N - 1)$ column matrix with element σ_j^2 for $j \neq i$ in row r; and Y is an $(N - 1)$ column matrix with element $\sum_{k=1, \neq j}^{N} V_{jk}$ in row r.

The inverse of B is derived to be an $(N - 1)$ square matrix with diagonal element $2/(2N - 3)$ and off-diagonal element $-1/(N - 2)(2N - 3)$. Therefore, upon premultiplying Y by B^{-1}, the element in row r is the ME of σ_j^2:

$$\tilde{\sigma}_j^2 = \frac{2 \displaystyle\sum_{k=1, \neq j}^{N} V_{jk}}{2N - 3} - \frac{\displaystyle\sum_{l \neq i, j} \sum_{k=1, \neq l}^{N} V_{lk}}{(N - 2)(2N - 3)}$$

$$= \frac{2(N - 2) \displaystyle\sum_{\substack{k=1 \\ \neq j}}^{N} V_{jk} - \displaystyle\sum_{l \neq i, j} \sum_{\substack{k=1 \\ \neq l}}^{N} V_{lk}}{(N - 2)(2N - 3)} \tag{7.B.2}$$

which is Eq. 7.2.7 in the text, thus completing the derivation.

DERIVATION OF MAXIMUM LIKELIHOOD ESTIMATORS FOR $N = 3$; $\sigma_1^2 = \sigma_2^2 = \sigma_0^2$; $\sigma_3^2 \neq \sigma_0^2$: EQUATIONS 7.2.10 AND 7.2.11

The likelihood function for $\sigma_1^2 \neq \sigma_2^2 \neq \sigma_3^2$ is given as Eq. 4.A.12. For $\sigma_1^2 = \sigma_2^2 = \sigma_0^2$ and $\sigma_3^2 \neq \sigma_0^2$, the likelihood function becomes

$$L = C - 0.5n \ln\left(\sigma_0^4 + 2\sigma_0^2\sigma_3^2\right) - \frac{0.5(n-1)}{\sigma_0^4 + 2\sigma_0^2\sigma_3^2}\left[\sigma_3^2 V_{12} + \sigma_0^2(V_{13} + V_{23})\right]$$

(7.C.1)

Then, $\partial L/\partial \sigma_0^2 = 0$ reduces to the following:

$$2n\left(\sigma_0^6 + 3\sigma_0^4\sigma_3^2 + 2\sigma_0^2\sigma_3^4\right) - (n-1)\left[2\sigma_3^2\left(\sigma_0^2 + \sigma_3^2\right)V_{12} + \sigma_0^4(V_{13} + V_{23})\right] = 0$$

(7.C.2)

and $\partial L/\partial \sigma_3^2 = 0$ simplifies to

$$2n\left(\sigma_0^2 + 2\sigma_3^2\right) - (n-1)(-V_{12} + 2V_{13} + 2V_{23}) = 0 \qquad (7.C.3)$$

Solving Eq. 7.C.3 for σ_0^2 gives the MLE of σ_0^2 as a function of σ_3^2:

$$\hat{\sigma}_0^2 = \frac{(n-1)(-V_{12} + 2V_{13} + 2V_{23})}{2n} - 2\sigma_3^2 \qquad (7.C.4)$$

Now, surmise that the MLE's of σ_0^2 and σ_3^2 are, respectively,

$$\hat{\sigma}_0^2 = \frac{(n-1)V_{12}}{2n} \qquad (7.C.5)$$

$$\hat{\sigma}_3^2 = \frac{(n-1)(V_{13} + V_{23} - V_{12})}{2n} \qquad (7.C.6)$$

and show that for these values of σ_0^2 and σ_3^2, Eqs. 7.C.2 and 7.C.4 are satisfied. Once this is shown, it will have been proven that Eqs. 7.C.5 and 7.C.6 give the MLE's of σ_0^2 and σ_3^2, respectively.

Simplify the notation slightly by writing $V_0 = V_{13} + V_{23}$. Then, the RHS of Eq. 7.C.4 is

$$\frac{(n-1)(-V_{12} + 2V_0)}{2n} - \frac{2(n-1)(V_0 - V_{12})}{2n} = \frac{(n-1)V_{12}}{2n} \qquad (7.C.7)$$

which is the same as the LHS of Eq. 7.C.4, thus showing that Eq. 7.C.4 is satisfied for the values of σ_0^2 and σ_3^2 given by Eqs. 7.C.5 and 7.C.6.

Finally, Eq. 7.C.2 becomes

$$\frac{(n-1)^3}{4n^2}\left(V_{12}^3 + 3V_0 V_{12}^2 - 3V_{12}^3 + 2V_0^2 V_{12} - 4V_0 V_{12}^2 + 2V_{12}^3 - 2V_0^2 V_{12}\right.$$

$$+ 2V_0 V_{12}^2 - V_0 V_{12}^2\Big) = 0$$

thus completing the derivation of the MLE's of σ_0^2 and σ_3^2.

DERIVATION OF EQUATION 7.4.27

Intermediate results are given to make the transition from Eqs. 7.4.23 and 7.4.26 to Eq. 7.4.27. Considerable algebra is involved. To simplify the notation, the expression $(\sigma_0^2 + \sigma_3^2)$, which appears often in this development, is labeled c so that, from Eq. 7.4.26,

$$V = \frac{2}{n-1} \begin{pmatrix} 4\sigma_0^4 & \sigma_0^4 & \sigma_0^4 \\ & c^2 & \sigma_3^4 \\ \text{symmetric} & & c^2 \end{pmatrix} \tag{7.D.1}$$

Then,

$$|V| = \frac{16\sigma_0^4}{(n-1)^3} \left(2c^4 - c^2\sigma_0^4 + \sigma_0^4\sigma_3^4 - 2\sigma_3^8 \right) \tag{7.D.2}$$

$$V^{-1} = \frac{4}{(n-1)^2 |V|} \begin{pmatrix} c_{11} & c_{12} & c_{13} \\ c_{12} & c_{22} & c_{23} \\ c_{13} & c_{23} & c_{33} \end{pmatrix} \tag{7.D.3}$$

where

$$c_{11} = c^4 - \sigma_3^8$$

$$c_{12} = c_{13} = \sigma_0^4 \left(\sigma_3^4 - c^2 \right)$$

$$c_{22} = c_{33} = \sigma_0^4 \left(4c^2 - \sigma_0^4 \right) \tag{7.D.4}$$

$$c_{23} = \sigma_0^4 \left(\sigma_0^4 - 4\sigma_3^4 \right)$$

After further algebra,

$$TV^{-1}T' = \frac{16}{(n-1)^2 |V|} \begin{pmatrix} b_{11} & b_{12} \\ b_{21} & b_{22} \end{pmatrix} \tag{7.D.5}$$

where

$$b_{11} = c^4 - \sigma_3^8$$

$$b_{12} = b_{21} = \sigma_0^4(c^2 - \sigma_3^4) \qquad (7.D.6)$$

$$b_{22} = 2\sigma_0^4(c^2 - \sigma_3^4)$$

$$(TV^{-1}T')^{-1} = \frac{(n-1)^2|V|\begin{pmatrix} a_{11} & a_{12} \\ a_{21} & a_{22} \end{pmatrix}}{16D\sigma_0^4} \qquad (7.D.7)$$

where

$$a_{11} = 2\sigma_0^4(c^2 - \sigma_3^4)$$

$$a_{12} = a_{21} = -\sigma_0^4(c^2 - \sigma_3^4) \qquad (7.D.8)$$

$$a_{22} = c^4 - \sigma_3^8$$

and

$$D = 2c^6 - (\sigma_0^4 + 2\sigma_3^4)c^4 + 2\sigma_3^4(\sigma_0^4 - \sigma_3^4)c^2 + \sigma_3^8(2\sigma_3^4 - \sigma_0^4) \qquad (7.D.9)$$

Upon premultiplying $(TV^{-1}T')^{-1}$ by $V^{-1}T'$, the result is

$$V^{-1}T'(TV^{-1}T')^{-1} = \begin{pmatrix} 0.5 & -0.5 \\ 0 & 0.5 \\ 0 & 0.5 \end{pmatrix} \qquad (7.D.10)$$

which is Eq. 7.4.27, thus completing the derivation.

REFERENCES

1. Grubbs, F. E. "On Estimating Precision of Measuring Instruments and Product Variability." *J. Amer. Stat. Assn.* **43**: 243–264; 1948.

2. Aitken, A. C. "On Least Squares and Linear Combinations of Observations." *Proc. Royal Soc. Edinburgh.* **55**: 42–48; 1935.

3. Browne, M. W. "Generalized Least Squares Estimators in the Analysis of Covariance Structures." *South African Stat. J.* **8**(1): 1–24; 1974.

4. Jaech, J. L. "Large Sample Tests for Grubbs' Estimators of Instrument Precision With More Than Two Instruments." *Technometrics.* **18**: 127–133; 1976.

Chapter 8

BIBLIOGRAPHY

At the end of each chapter, papers referenced in that chapter are listed. A number of additional references bear rather directly on the subject of this monograph. They are listed here for readers who wish to pursue the subject further. Some additional relevant references are found in the sources listed below.

1. *ASTM Manual for Conducting an Interlaboratory Study of a Test Method*. ASTM STP 335. American Society for Testing and Materials: 1963.
2. Gaylor, D. W. "Equivalence of Two Estimates of Product Variance." *J. Amer. Stat. Assn.* **51**: 451–453; 1956.
3. Grubbs, F. E. "Errors of Measurement, Precision, Accuracy and the Statistical Comparison of Measuring Instruments." *Technometrics*. **15**: 53–66; 1973.
4. Hahn, G. J. and Nelson, W. "A Problem in the Statistical Comparison of Measuring Devices." *Technometrics*. **12**: 95–102; 1970.
5. Hanamura, R. C. "Estimating Imprecisions of Measuring Instruments." *Technometrics*. **17**: 299–302; 1975.
6. Hanamura, R. C. "Estimating Variances in Simultaneous Measurement Procedures." *Amer. Statistician*. **29**: 108–109; 1975.
7. Hartung, J. "Non-Negative Minimum Biased Invariant Estimation in Variance Component Models." *Ann. Stat.* **9**: 278–292; 1981.
8. Hemmerle, W. J. and Downs, B. W. "Nonhomogenous Variances in the Mixed AOV Model; Maximum Likelihood Estimation." *Contributions to Survey Sampling and Applied Statistics*. New York: Academic Press; 1978.
9. Jaech, J. L. "Case Studies in the Statistical Analysis of Safeguards Data." *IAEA-SM-201/14. Proceedings of IAEA Symposium on Safeguards, Vienna*; pp. 545–559; 1976.
10. Jaech, J. L. "Errors of Measurement With More Than Two Measurement Methods." *Nucl. Mat. Manage. J.* **IV**: Winter 1976.
11. Jaech, J. L. "Extension of Grubbs' Method When Relative Biases are Not Constant." *Nucl. Mat. Manage. J.* **VIII**: 76–80; 1979.

12. Jaech, J. L. "Making Inferences About the Shipper's Variance in a Shipper-Receiver Difference Situation." *Nucl. Mat. Manage. J.* **IV**(1): Spring 1975.

13. Jaech, J. L. "Pitfalls in the Analysis of Paired Data." *INMM J.* **II**(2):32–39; 1973.

14. Mandel, J. "A New Analysis of Variance Model for Non-Additive Data." *Technometrics*. **13**: 1–18; 1971.

15. Mandel, J. "The Measuring Process." *Technometrics*. **1**(3): 251–267; 1959.

16. Mandel, J. "Repeatability and Reproducibility." *J. Qual. Technol.* **4**: 74–85; 1972.

17. Mandel, J. and Lashof, T. W. "The Interlaboratory Evaluation of Testing Methods." *ASTM Bulletin No. 239, Technical Publication 133. American Society of Testing and Materials*: July 1959, pp. 53–61.

18. Maxwell, E. A. "Estimating Variances from One or Two Measurements on Each Sample." *Amer. Statistician*. **28**: 96–97; 1974.

19. Salsburg, D. S. "Estimating Differences in Variance When Comparing Two Methods of Assay." *Technometrics*. **17**: 381–382; 1975.

20. Shukla, G. K. "An Invariant Test for the Homogeneity of Variances in a Two-Way Classification." *Biometrics*. **28**: 1063–1072; 1972.

21. Shukla, G. K. "Some Exact Tests of Hypotheses About Grubbs' Estimators." *Biometrics*. **29**: 373–378; 1973.

22. Smith, H. F. "Estimating Precision of Measuring Instruments." *J. Amer. Stat. Assn*. **45**: 447–451; September 1950.

23. Winslow, G. H. "A Problem Related to Grubbs' Technique: The Distribution of Method Variance Estimates for Known Product Variance." *INMM J.* **V**(4): 26–29; 1976.

Chapter 9

COMPUTER PROGRAMS

The program listings that provide the maximum likelihood estimators (MLE's) for the paired data (Chapters 4 and 5) and for the original data (Chapter 6) are given here, identified as Programs 1 and 2, respectively.

The program input and output are illustrated for an example for which the data were generated using random normal deviates and the following parameter values:

$$\sigma_\mu^2 = 1$$

$$\sigma_1^2 = 1.44 \qquad \sigma_3^2 = 1$$

$$\sigma_2^2 = 0.25 \qquad \sigma_4^2 = 1$$

$$\beta_i = 1 \qquad \text{for all } i$$

Section 9.A relates to the paired data and utilizes Program 1. The data are listed first. The R column gives the MLE's of σ_1^2, σ_2^2, σ_3^2, and σ_4^2, respectively, for each iteration. The columns labeled B_0, B_1, and B_2 are the respective b_{0i}, b_{1i}, and b_{2i} values. Note that the final MLE's are

$$\hat{\sigma}_1^2 = 1.2669 \qquad \hat{\sigma}_3^2 = 0.8887$$

$$\hat{\sigma}_2^2 = 0.1961 \qquad \hat{\sigma}_4^2 = 1.4592$$

The respective moment estimators (ME's) are in good agreement in this example:

$$\tilde{\sigma}_1^2 = 1.5332 \qquad \tilde{\sigma}_3^2 = 0.8762$$

$$\tilde{\sigma}_2^2 = 0.2136 \qquad \tilde{\sigma}_4^2 = 1.3826$$

Section 9.B gives the results for the original data, utilizing Program 2 and for the constant bias model ($\beta_i = 1$ for all i). $R1$, $R2$, ... are the MLE's of $\sigma_1^2, \sigma_2^2, \ldots$ after each iteration. Also, $R0$ is the MLE of σ_μ^2, CI is $a_i = \beta_i \bar{\sigma}_\mu$, DI is a_i^2 or b_{0i}, EI is b_{1i}, FI is b_{2i}, and GI is b_{3i}. The final MLE's are

$$\hat{\sigma}_1^2 = 1.2055 \qquad \hat{\sigma}_3^2 = 0.8384$$

$$\hat{\sigma}_2^2 = 0.2615 \qquad \hat{\sigma}_4^2 = 1.3791$$

$$\hat{\sigma}_\mu^2 = 1.3658$$

Section 9.C also relates to the original data utilizing Program 2, but for the nonconstant bias model. The program output is as just described for the constant bias model. Note that the values for β_i range from 0.63 to 1.41, apparently quite different from unity. However, a test of the hypothesis $\beta_i = 1$ for all i is not rejected, and hence, the parameter estimates for the constant bias model are the appropriate ones to use. For the nonconstant bias model, the MLE's are

$$\hat{\sigma}_1^2 = 1.2013 \qquad \hat{\sigma}_3^2 = 0.9559$$

$$\hat{\sigma}_2^2 = 0.0556 \qquad \hat{\sigma}_4^2 = 1.5419$$

$$\hat{\sigma}_\mu^2 = 0.6441$$

and the corresponding ME's are

$$\tilde{\sigma}_1^2 = 1.4120 \qquad \tilde{\sigma}_3^2 = 0.9794$$

$$\tilde{\sigma}_2^2 = 0.0002 \qquad \tilde{\sigma}_4^2 = 1.4036$$

$$\tilde{\sigma}_\mu^2 = 0.6029$$

Sections 9.D and 9.E list Programs 1 and 2, respectively.

9.A RESULTS FOR PAIRED DATA

FILE: CONSOUT CRTCOPY A3 EXXON NUCLEAR - VM/CMS

PLEASE ENTER INPUT DATA FILENAME

STA21
DO YOU WANT TO SEE THE INPUT DATA? (Y OR N)
Y

INPUT DATA:

```
   9.920 10.330 11.660 11.390

  10.800 10.270 10.230  8.740

   7.390  0.610 11.470  9.770

  10.160  9.780 11.170 11.380

  10.320 10.910 11.420 11.710

  13.610 12.170 13.080 10.450

   9.210 10.530  9.840  9.410

   8.780  8.900  8.670  7.640

   9.770  8.790  8.510  9.140

  10.410 11.570 10.490  8.090

   9.880 11.710 11.270 10.420

  10.760 11.010  9.940 13.090

  10.130 10.590  9.600  9.990

   9.440 10.180  9.650  9.200

  10.120  8.750 10.440  8.450

  12.270  9.510  8.730  9.210

   8.660  9.960 10.590  9.800

   9.930  9.610  9.890 10.620

  10.210  8.390  7.610  8.350

  10.920 11.110 10.620 12.140
```

DO YOU WANT TO SEE THE MEANS AND VARIANCES? (Y OR N)
Y

MEANS (M) AND VARIANCES (V):

M 1 2 =   0.00050   V 1 2 =   1.440646

```
FILE: CONSOUT   CRTCOPY   A3   EXXON NUCLEAR - VM/CMS


M  1  3 =   -0.10950      V  1  3 =    2.653319
M  1  4 =    0.18500      V  1  4 =    2.978066
M  2  3 =   -0.11000      V  2  3 =    1.152105
M  2  4 =    0.18450      V  2  4 =    1.840088
M  3  4 =    0.29450      V  3  4 =    1.952699
PLEASE ENTER VALUES FOR R2 ETC.
VALUE FOR R   2
?
.25
VALUE FOR R   3
?
1
VALUE FOR R   4
?
1
BEGINNING VALUES FOR R:

  R   2 =   0.25000
  R   3 =   1.00000
  R   4 =   1.00000

CORRECT? (Y OR N)
Y


R        R
NUM      VALUE             B0              B1              B2
  1  1.27000654         6.0000       11.393970        13.9215
  2  0.26672406         2.7874        4.126554         6.3868
  3  0.83579731         5.5366        8.361379        13.4967
  4  1.39991156         5.7331       11.580098        11.9207


ANOTHER ITERATION USING COMPUTED R VALUES? (Y OR N)
Y


R        R
NUM      VALUE             B0              B1              B2
  1  1.27090762         5.6600       10.703182        11.7650
  2  0.21435883         2.6976        3.826438         5.8407
  3  0.87526397         6.1662        8.857264        13.0939
  4  1.44218934         6.5944       13.158391        13.8140


ANOTHER ITERATION USING COMPUTED R VALUES? (Y OR N)
Y


R        R
NUM      VALUE             B0              B1              B2
  1  1.26643736         6.5010       11.817135        13.6397
  2  0.20088800         2.6255        3.729751         5.5712
  3  0.88512071         6.4609        9.184147        13.6445
  4  1.45464943         6.8973       13.717440        14.5091


ANOTHER ITERATION USING COMPUTED R VALUES? (Y OR N)
```

```
FILE: CONSOUT  CRTCOPY  A3  EXXON NUCLEAR - VM/CMS

Y

R       R
NUM     VALUE            B0              B1              B2
  1 1.26668585          6.7951          12.216356       14.2929
  2 0.19737840          2.6067          3.703941        5.4994
  3 0.88772657          6.5433          9.274114        13.7873
  4 1.45797103          6.9823          13.873373       14.6971


ANOTHER ITERATION USING COMPUTED R VALUES? (Y OR N)
Y

R       R
NUM     VALUE            B0              B1              B2
  1 1.26684810          6.8788          12.330408       14.4782
  2 0.19645166          2.6017          3.697092        5.4804
  3 0.88841579          6.5656          9.298323        13.8254
  4 1.45885381          7.0053          13.915344       14.7473


ANOTHER ITERATION USING COMPUTED R VALUES? (Y OR N)
Y

R       R
NUM     VALUE            B0              B1              B2
  1 1.26689803          6.9014          12.361285       14.5283
  2 0.19620632          2.6004          3.695277        5.4753
  3 0.88859837          6.5715          9.304763        13.8355
  4 1.45908792          7.0114          13.926512       14.7606


ANOTHER ITERATION USING COMPUTED R VALUES? (Y OR N)
Y

R       R
NUM     VALUE            B0              B1              B2
  1 1.26691174          6.9074          12.369513       14.5417
  2 0.19614131          2.6000          3.694796        5.4740
  3 0.88864675          6.5730          9.306472        13.8382
  4 1.45914998          7.0130          13.929475       14.7641


ANOTHER ITERATION USING COMPUTED R VALUES? (Y OR N)
Y

R       R
NUM     VALUE            B0              B1              B2
  1 1.26691541          6.9090          12.371697       14.5452
  2 0.19612409          2.6000          3.694668        5.4736
  3 0.88865957          6.5735          9.306925        13.8389
  4 1.45916643          7.0134          13.930261       14.7651


ANOTHER ITERATION USING COMPUTED R VALUES? (Y OR N)
```

9.B RESULTS FOR ORIGINAL DATA AND CONSTANT BIAS MODEL

```
FILE: CONSOUT  CRTCOPY  A3  EXXON NUCLEAR - VM/CMS


PLEASE ENTER INPUT DATA FILENAME

STA21

DO YOU WANT TO SEE THE INPUT DATA? (Y OR N)
N

DO YOU WANT TO SEE THE MEAN AND VARIANCE? (Y OR N)
Y

MEAN (M) AND VARIANCE (V):

M  1 =   10.13450    V  1  1 =   1.651110
M  2 =   10.13400    V  2  2 =   1.203436
M  3 =   10.24400    V  3  3 =   1.635120
M  4 =    9.94950    V  4  4 =   2.103984

DO YOU WANT TO SEE THE COVARIANCE? (Y OR N)
Y

COVARIANCES (C) :

C  1  2 =    0.706949
C  1  3 =    0.316455
C  1  4 =    0.388513
C  2  3 =    0.843225
C  2  4 =    0.733665
C  3  4 =    0.893202

DO YOU WANT ALL BS SET TO 1.0 ? (Y OR N)
Y

BS ARE SET TO 1.0

B  1 =    1.0000     A  1 =    0.0190
B  2 =    1.0000     A  2 =    0.0185
B  3 =    1.0000     A  3 =    0.1285
B  4 =    1.0000     A  4 =   -0.1660

VALUE FOR R   1
?
1.44
VALUE FOR R   2
?
.25
VALUE FOR R   3
?
1
VALUE FOR R   4
?
1

BEGINNING VALUES FOR R:
```

```
FILE: CONSOUT   CRTCOPY   A3   EXXON NUCLEAR - VM/CMS


   R   1 =   1.44000
   R   2 =   0.25000
   R   3 =   1.00000
   R   4 =   1.00000

CORRECT? (Y OR N)
Y


RO =    1.376437

   C  1 =    1.173216     D  1 =    1.376437
   C  2 =    1.173216     D  2 =    1.376437
   C  3 =    1.173216     D  3 =    1.376437
   C  4 =    1.173216     D  4 =    1.376437

   R 1 =        1.20717992
   E 1 =           9.2586
   F 1 =          17.334199
   G 1 =          27.7149

   R 2 =        0.29450520
   E 2 =           4.8931
   F 2 =           6.964640
   G 2 =          14.2156

   R 3 =        0.81894777
   E 3 =           8.1904
   F 3 =          12.732851
   G 3 =          25.1313

   R 4 =        1.35966006
   E 4 =           8.4947
   F 4 =          17.381652
   G 4 =          23.2975



ANOTHER ITERATION USING COMPUTED R VALUES? (Y OR N)
Y


RO =    1.383098

   C  1 =    1.176052     D  1 =    1.383098
   C  2 =    1.176052     D  2 =    1.383098
   C  3 =    1.176052     D  3 =    1.383098
   C  4 =    1.176052     D  4 =    1.383098

   R 1 =        1.20778977
   E 1 =           8.4025
   F 1 =          15.927406
   G 1 =          23.0185

   R 2 =        0.26442733
```

```
FILE: CONSOUT   CRTCOPY   A3   EXXON NUCLEAR - VM/CMS


E 2 =        4.8513
F 2 =        6.670756
G 2 =       13.5550

R 3 =        0.83669128
E 3 =        8.3929
F 3 =       12.686056
G 3 =       23.2915

R 4 =        1.37729033
E 4 =        9.0287
F 4 =       18.366894
G 4 =       24.9452



ANOTHER ITERATION USING COMPUTED R VALUES? (Y OR N)
     SENT FILE 3526 (3526) ON LINK BELLPRT TO BELLPRT SYSTEM
Y


RO =    1.367340

    C  1 =   1.169333      D  1 =   1.367340
    C  2 =   1.169333      D  2 =   1.367340
    C  3 =   1.169333      D  3 =   1.367340
    C  4 =   1.169333      D  4 =   1.367340

R 1 =        1.20560938
E 1 =        8.7979
F 1 =       16.393289
G 1 =       24.3787

R 2 =        0.26174175
E 2 =        4.7611
F 2 =        6.546934
G 2 =       13.2173

R 3 =        0.83825901
E 3 =        8.3509
F 3 =       12.601572
G 3 =       23.1691

R 4 =        1.37892607
E 4 =        8.9893
F 4 =       18.279364
G 4 =       24.9302



ANOTHER ITERATION USING COMPUTED R VALUES? (Y OR N)
Y


RO =    1.365916
```

```
FILE: CONSOUT  CRTCOPY  A3  EXXON NUCLEAR - VM/CMS


    C  1 =    1.168724      D  1 =    1.365916
    C  2 =    1.168724      D  2 =    1.365916
    C  3 =    1.168724      D  3 =    1.365916
    C  4 =    1.168724      D  4 =    1.365916

    R 1 =     1.20551776
    E 1 =        8.8386
    F 1 =       16.442687
    G 1 =       24.5180

    R 2 =     0.26150791
    E 2 =        4.7531
    F 2 =        6.535811
    G 2 =       13.1870

    R 3 =     0.83839003
    E 3 =        8.3468
    F 3 =       12.593459
    G 3 =       23.1565

    R 4 =     1.37906389
    E 4 =        8.9855
    F 4 =       18.270865
    G 4 =       24.9274


ANOTHER ITERATION USING COMPUTED R VALUES? (Y OR N)
Y


RO =    1.365790

    C  1 =    1.168670      D  1 =    1.365790
    C  2 =    1.168670      D  2 =    1.365790
    C  3 =    1.168670      D  3 =    1.365790
    C  4 =    1.168670      D  4 =    1.365790

    R 1 =     1.20551156
    E 1 =        8.8422
    F 1 =       16.447073
    G 1 =       24.5303

    R 2 =     0.26148867
    E 2 =        4.7524
    F 2 =        6.534852
    G 2 =       13.1844

    R 3 =     0.83840077
    E 3 =        8.3465
    F 3 =       12.592712
    G 3 =       23.1553

    R 4 =     1.37907508
```

```
FILE: CONSOUT  CRTCOPY  A3  EXXON NUCLEAR - VM/CMS


E 4 =           8.9851
F 4 =         18.270056
G 4 =          24.9270



ANOTHER ITERATION USING COMPUTED R VALUES? (Y OR N)
Y


R0 =    1.365779

  C  1 =     1.168666     D  1 =    1.365779
  C  2 =     1.168666     D  2 =    1.365779
  C  3 =     1.168666     D  3 =    1.365779
  C  4 =     1.168666     D  4 =    1.365779

  R 1 =     1.20551107
  E 1 =        8.8425
  F 1 =       16.447435
  G 1 =       24.5313

  R 2 =     0.26148711
  E 2 =        4.7523
  F 2 =        6.534773
  G 2 =       13.1842

  R 3 =     0.83840164
  E 3 =        8.3464
  F 3 =       12.592650
  G 3 =       23.1552

  R 4 =     1.37907599
  E 4 =        8.9851
  F 4 =       18.269988
  G 4 =       24.9270



ANOTHER ITERATION USING COMPUTED R VALUES? (Y OR N)
N
ANOTHER ITERATION USING NEW R VALUES? (Y OR N)
N
ENC CCS CRTCOPY EXEC (VERSION 1.0 3/82)
```

9.C RESULTS FOR ORIGINAL DATA AND NONCONSTANT BIAS MODEL

```
FILE: CONSOUT  CRTCOPY  A3  EXXON NUCLEAR - VM/CMS


PLEASE ENTER INPUT DATA FILENAME

STA21

DO YOU WANT TO SEE THE INPUT DATA? (Y OR N)
N

DO YOU WANT TO SEE THE MEAN AND VARIANCE? (Y OR N)
N

DO YOU WANT TO SEE THE COVARIANCE? (Y OR N)
N

DO YOU WANT ALL BS SET TO 1.0 ? (Y OR N)
N

BS WILL BE CALCULATED

B  1 =    0.6298      A  1 =    3.7641
B  2 =    1.4127      A  2 =   -4.1559
B  3 =    1.0429      A  3 =   -0.3051
B  4 =    1.0778      A  4 =   -0.9533

VALUE FOR R   1
?
1.44
VALUE FOR R   2
?
.25
VALUE FOR R   3
?
1
VALUE FOR R   4
?
1

BEGINNING VALUES FOR R:

   R  1 =   1.44000
   R  2 =   0.25000
   R  3 =   1.00000
   R  4 =   1.00000

CORRECT? (Y OR N)
Y


R0 =    0.859007

   C  1 =    0.583683    D  1 =    0.340686
   C  2 =    1.309300    D  2 =    1.714266
   C  3 =    0.966555    D  3 =    0.934229
   C  4 =    0.998966    D  4 =    0.997933
```

```
FILE: CONSOUT  CRTCOPY  A3  EXXON NUCLEAR - VM/CMS


R 1 =      1.22001038
E 1 =         9.7892
F 1 =        13.944696
G 1 =        29.1519

R 2 =      0.15806506
E 2 =         3.2114
F 2 =         7.655481
G 2 =         9.9599

R 3 =      0.89033827
E 3 =        13.1225
F 3 =        16.279893
G 3 =        42.1125

R 4 =      1.45986525
E 4 =        13.1739
F 4 =        24.048233
G 4 =        36.3493


ANOTHER ITERATION USING COMPUTED R VALUES? (Y OR N)
Y


R0 =    0.762248

C  1 =    0.549828      D  1 =    0.302310
C  2 =    1.233357      D  2 =    1.521170
C  3 =    0.910492      D  3 =    0.828996
C  4 =    0.941023      D  4 =    0.885525

R 1 =      1.21058306
E 1 =        12.1614
F 1 =        16.675123
G 1 =        34.5835

R 2 =      0.06507265
E 2 =         2.7874
F 2 =         6.329512
G 2 =         8.3143

R 3 =      0.94922726
E 3 =        25.2328
F 3 =        28.501958
G 3 =        73.5702

R 4 =      1.53408206
E 4 =        25.4995
F 4 =        44.644987
G 4 =        73.0151
```

```
FILE: CONSOUT  CRTCOPY  A3  EXXON NUCLEAR - VM/CMS


ANOTHER ITERATION USING COMPUTED R VALUES? (Y OR N)
Y


R0 =    0.655501

   C  1 =    0.509877     D  1 =    0.259974
   C  2 =    1.143740     D  2 =    1.308142
   C  3 =    0.844335     D  3 =    0.712902
   C  4 =    0.872648     D  4 =    0.761514

   R 1 =    1.20208078
   E 1 =      22.3502
   F 1 =      29.331499
   G 1 =      66.8015

   R 2 =    0.05553369
   E 2 =       2.4637
   F 2 =       5.452888
   G 2 =       7.4051

   R 3 =    0.95583092
   E 3 =      25.2685
   F 3 =      28.311710
   G 3 =      75.7697

   R 4 =    1.54212577
   E 4 =      25.5179
   F 4 =      44.503968
   G 4 =      76.3795



ANOTHER ITERATION USING COMPUTED R VALUES? (Y OR N)
Y


R0 =    0.643954

   C  1 =    0.505366     D  1 =    0.255395
   C  2 =    1.133621     D  2 =    1.285098
   C  3 =    0.836865     D  3 =    0.700344
   C  4 =    0.864927     D  4 =    0.748099

   R 1 =    1.20128104
   E 1 =      25.3587
   F 1 =      33.103551
   G 1 =      76.3061

   R 2 =    0.05562331
   E 2 =       2.4304
   F 2 =       5.361779
   G 2 =       7.3129
        SENT FILE 3530 (3530) ON LINK BELLPRT TO BELLPRT SYSTEM
```

FILE: CONSOUT   CRTCOPY   A3   EXXON NUCLEAR - VM/CMS


```
R 3 =      0.95585975
E 3 =        24.8013
F 3 =        27.798723
G 3 =        74.6241

R 4 =       1.54193134
E 4 =        25.0489
F 4 =        43.694968
G 4 =        75.3723
```


ANOTHER ITERATION USING COMPUTED R VALUES? (Y OR N)
Y


```
R0 =     0.644080

C  1 =     0.505415      D  1 =     0.255445
C  2 =     1.133733      D  2 =     1.285350
C  3 =     0.836948      D  3 =     0.700481
C  4 =     0.865012      D  4 =     0.748247

R 1 =       1.20128781
E 1 =        25.3262
F 1 =        33.062830
G 1 =        76.2024

R 2 =       0.05561318
E 2 =         2.4307
F 2 =         5.362725
G 2 =         7.3137

R 3 =       0.95586413
E 3 =        24.8102
F 3 =        27.808383
G 3 =        74.6502

R 4 =       1.54194221
E 4 =        25.0578
F 4 =        43.710231
G 4 =        75.3949
```


ANOTHER ITERATION USING COMPUTED R VALUES? (Y OR N)
Y


```
R0 =     0.644068

C  1 =     0.505410      D  1 =     0.255440
C  2 =     1.133722      D  2 =     1.285325
C  3 =     0.836939      D  3 =     0.700468
C  4 =     0.865004      D  4 =     0.748232
```

```
FILE: CONSOUT   CRTCOPY   A3   EXXON NUCLEAR - VM/CMS


       R 1 =      1.20128699
       E 1 =        25.3299
       F 1 =       33.067503
       G 1 =        76.2142

       R 2 =      0.05561375
       E 2 =         2.4307
       F 2 =        5.362628
       G 2 =         7.3136

       R 3 =      0.95586389
       E 3 =        24.8095
       F 3 =       27.807612
       G 3 =        74.6483

       R 4 =      1.54194154
       E 4 =        25.0571
       F 4 =       43.709011
       G 4 =        75.3932



ANOTHER ITERATION USING COMPUTED R VALUES? (Y OR N)
N
ANOTHER ITERATION USING NEW R VALUES? (Y OR N)
N
ENC CCS CRTCOPY EXEC (VERSION 1.0 3/82)
```

9.D   PROGRAM 1 — PAIRED DATA

FILE: STARPT05 FORTRAN   A   EXXON NUCLEAR - VM/CMS

```
C                                                                      STA00010
C      FORTRAN PROGRAM WRITTEN FOR JOHN JAECH PER HIS INSTRUCTIONS      STA00020
C      RE PUBLICATION :   SPECIAL PROBLEMS IN MEASUREMENT VARIANCE      STA00030
C       ESTIMATION AND HYPOTHESIS TESTING                              STA00040
C      PROGRAM 1 ESTIMATION OF MEASUREMENT PRECISIONS FROM NI BY N ARRAY:STA00050
C       CONSTANT BIAS MODEL (SEC. 7.2)                                 STA00060
C                                                                      STA00070
C      INPUT OF NI BY N ARRAY IS FROM A FILE (01)                      STA00080
C        DIMENSIONS OF THE ARRAY IS THE FIRST RECORD IN THAT FILE      STA00090
C        MAX ARRAY SIZE (20,500)                                       STA00100
C                                                                      STA00110
       CHARACTER*1 AY,AN,RESP                                          STA00120
       DOUBLE PRECISION X(20,500),AVE(20,20),VAR(20,20),R(20),SUM,     STA00130
      1 SUMSQ,SQSUM,TEMP,B0,B1,B2                                      STA00140
       DATA AY/'Y'/,AN/'N'/                                            STA00150
C                                                                      STA00160
C      READ IN DIMENSIONS (N LT 21 AND NI LT 501)                      STA00170
C                                                                      STA00180
       READ (1,*) N,NI                                                 STA00190
       IF (N.LE.20.AND.NI.LE.500.AND.N.GT.0.AND.NI.GT.0) GO TO 100     STA00200
       WRITE (6,1000)                                                  STA00210
       STOP                                                            STA00220
C                                                                      STA00230
C      READ IN DATA POINTS FROM FILE 01 (THERE ARE 10 DATA ELEMENTS    STA00240
C      PER RECORD; SKIP ONE RECORD BETWEEN THE EACH 'ROW'              STA00250
C                                                                      STA00260
   100 DO 110 I=1,NI                                                   STA00270
       IF (N.LE.10) THEN                                               STA00280
         READ (1,*,END=500) (X(J,I),J=1,N)                             STA00290
       ELSE                                                            STA00300
         READ (1,*,END=500) (X(J,I),J=1,10)                            STA00310
         READ (1,*,END=500) (X(J,I),J=11,N)                            STA00320
       END IF                                                          STA00330
       READ (1,1010)                                                   STA00340
   110 CONTINUE                                                        STA00350
C                                                                      STA00360
C      CHECK IF WANT INPUT DATA PRINTED OUT                            STA00370
C                                                                      STA00380
   115 WRITE (6,1005)                                                  STA00390
       READ (5,1070) RESP                                              STA00400
       IF (RESP.EQ.AN) GO TO 130                                       STA00410
       IF (RESP.EQ.AY) GO TO 120                                       STA00420
       WRITE (6,1080)                                                  STA00430
       GO TO 115                                                       STA00440
C                                                                      STA00450
C      PRINT OUT INPUT DATA                                            STA00460
C                                                                      STA00470
   120 WRITE (6,1006)                                                  STA00480
       DO 125 I=1,NI                                                   STA00490
       IF (N.LE.10) THEN                                               STA00500
         WRITE (6,1007) (X(J,I),J=1,N)                                 STA00510
       ELSE                                                            STA00520
         WRITE (6,1007) (X(J,I),J=1,10)                                STA00530
         WRITE (6,1007) (X(J,I),J=11,N)                                STA00540
       END IF                                                          STA00550
```

```
FILE: STARPT05 FORTRAN   A    EXXON NUCLEAR - VM/CMS


        WRITE (6,1010)                                             STA00560
    125 CONTINUE                                                   STA00570
C                                                                  STA00580
C       CLEAR MEAN AND VARIANCE ARRAYS                             STA00590
C                                                                  STA00600
    130 DO 140 I=1,20                                              STA00610
        DO 140 J=1,20                                              STA00620
        VAR(I,J)=0.0                                               STA00630
        AVE(I,J)=0.0                                               STA00640
    140 CONTINUE                                                   STA00650
C                                                                  STA00660
C       CALCULATE THE MEAN AND VARIANCE OF THE COLUMNS OF DIFFERENCE STA00670
C                                                                  STA00680
        DO 180 I=1,N                                               STA00690
        DO 180 J=1,N                                               STA00700
        SUM=0.0                                                    STA00710
        SUMSQ=0.0                                                  STA00720
        DO 160 K=1,NI                                              STA00730
        IF (J.EQ.I) GO TO 170                                      STA00740
        IF (J.GI.I) GO TO 150                                      STA00750
        VAR(I,J)=VAR(J,I)                                          STA00760
        GO TO 170                                                  STA00770
    150 DIF=X(I,K)-X(J,K)                                          STA00780
        SUM=SUM+DIF                                                STA00790
        SUMSQ=SUMSQ+(DIF**2)                                       STA00800
    160 CONTINUE                                                   STA00810
        AVE(I,J)=SUM/NI                                            STA00820
        SQSUM=SUM**2                                               STA00830
        TEMP=SUMSQ-SQSUM/NI                                        STA00840
        VAR(I,J)=TEMP/(NI-1)                                       STA00850
    170 CONTINUE                                                   STA00860
    180 CONTINUE                                                   STA00870
C                                                                  STA00880
C       CHECK IF WANT MEAN AND VARIANCE DATA PRINTED OUT           STA00890
C                                                                  STA00900
    185 WRITE (6,1012)                                             STA00910
        READ (5,1070) RESP                                         STA00920
        IF (RESP.EQ.AN) GO TO 210                                  STA00930
        IF (RESP.EQ.AY) GO TO 190                                  STA00940
        WRITE (6,1080)                                             STA00950
        GO TO 185                                                  STA00960
C                                                                  STA00970
C       PRINT OUT MEAN AND VARIANCE DATA                          STA00980
C                                                                  STA00990
    190 WRITE (6,1014)                                             STA01000
        NLESS1=N-1                                                 STA01010
        DO 200 I=1,NLESS1                                          STA01020
        IPLUS1=I+1                                                 STA01030
        DO 200 J=IPLUS1,N                                          STA01040
        WRITE (6,1016) I,J,AVE(I,J),I,J,VAR(I,J)                   STA01050
    200 CONTINUE                                                   STA01060
C                                                                  STA01070
C       GET INITIAL VALUES FOR R2 THROUGH RN                       STA01080
C                                                                  STA01090
    210 R(1)=0.0                                                   STA01100
```

FILE: STARPT05 FORTRAN   A   EXXON NUCLEAR - VM/CMS

```
      WRITE (6,1020)                                          STA01110
      DO 220 I=2,N                                            STA01120
      WRITE (6,1030) I                                        STA01130
      READ (5,*) R(I)                                         STA01140
  220 CONTINUE                                                STA01150
C                                                             STA01160
C     ECHO BACK INPUTTED R VALUES FOR USER VERIFICATION       STA01170
C                                                             STA01180
  230 WRITE (6,1040)                                          STA01190
      DO 240 I=2,N                                            STA01200
      WRITE (6,1050) I,R(I)                                   STA01210
  240 CONTINUE                                                STA01220
  250 WRITE (6,1060)                                          STA01230
      READ (5,1070) RESP                                      STA01240
      IF (RESP.EQ.AY) GO TO 310                               STA01250
      IF (RESP.EQ.AN) GO TO 270                               STA01260
      WRITE (6,1080)                                          STA01270
      GO TO 250                                               STA01280
C                                                             STA01290
C     ALLOW USER TO CORRECT INPUTTED R VALUES                 STA01300
C                                                             STA01310
  270 WRITE (6,1090)                                          STA01320
      READ (5,*) I                                            STA01330
      IF (I.LE.N.AND.I.GT.1) GO TO 280                        STA01340
      WRITE (6,1085)                                          STA01350
      GO TO 270                                               STA01360
  280 WRITE (6,1110)                                          STA01370
      READ (5,*) R(I)                                         STA01380
  290 WRITE (6,1120)                                          STA01390
      READ (5,1070) RESP                                      STA01400
      IF (RESP.EQ.AY) GO TO 270                               STA01410
      IF (RESP.EQ.AN) GO TO 230                               STA01420
      WRITE (6,1080)                                          STA01430
      GO TO 290                                               STA01440
C                                                             STA01450
C     COMPUTE VALUES FOR B                                    STA01460
C                                                             STA01470
  310 WRITE (6,1130)                                          STA01480
      DO 400 I=1,N                                            STA01490
      B0=0.0                                                  STA01500
      B1=0.0                                                  STA01510
      B2=0.0                                                  STA01520
      DO 380 J=1,N                                            STA01530
      IF (I.EQ.J) GO TO 360                                   STA01540
      B0=B0+(1/R(J))                                          STA01550
      B1=B1+(VAR(I,J)/R(J))                                   STA01560
      JPLUS1=J+1                                              STA01570
      IF (JPLUS1.GT.N) GO TO 360                              STA01580
      DO 340 K=JPLUS1,N                                       STA01590
      IF (K.EQ.I) GO TO 320                                   STA01600
      B2=B2+(VAR(J,K)/(R(J)*R(K)))                            STA01610
  320 CONTINUE                                                STA01620
  340 CONTINUE                                                STA01630
  360 CONTINUE                                                STA01640
  380 CONTINUE                                                STA01650
```

```
FILE: STARPT05 FORTRAN  A    EXXON NUCLEAR - VM/CMS


      R(I)=(((NI-1)*(B0*B1-B2))/(NI*(B0**2)))-(1/B0)            STA01660
      WRITE (6,1140) I,R(I),B0,B1,B2                            STA01670
  400 CONTINUE                                                  STA01680
C                                                               STA01690
C     CHECK IF WANT MORE ITERATIONS                            STA01700
C                                                               STA01710
                                                                STA01720
  410 WRITE (6,1150)                                            STA01730
      READ (5,1070) RESP                                        STA01740
      IF (RESP.EQ.AY) GO TO 310                                 STA01750
      IF (RESP.EQ.AN) GO TO 420                                 STA01760
      WRITE (6,1080)                                            STA01770
      GO TO 410                                                 STA01780
  420 WRITE (6,1160)                                            STA01790
      READ (5,1070) RESP                                        STA01800
      IF (RESP.EQ.AY) GO TO 210                                 STA01810
      IF (RESP.EQ.AN) GO TO 440                                 STA01820
      WRITE (6,1080)                                            STA01830
      GO TO 420                                                 STA01840
  440 STOP                                                      STA01850
C                                                               STA01860
C     ERRORS                                                    STA01870
C                                                               STA01880
  500 WRITE (6,1170)                                            STA01890
      STOP                                                      STA01900
C                                                               STA01910
C     FORMATS                                                   STA01920
C                                                               STA01930
 1000 FORMAT (1X,'ERROR ON ARRAY DIMENSIONS')                  STA01940
 1005 FORMAT (1X,'DO YOU WANT TO SEE THE INPUT DATA? (Y OR N)') STA01950
 1006 FORMAT (/,1X,'INPUT DATA:',/)                            STA01960
 1007 FORMAT (1X,10F7.3)                                        STA01970
 1010 FORMAT (1X)                                               STA01980
 1012 FORMAT (1X,'DO YOU WANT TO SEE THE MEANS AND VARIANCES? (Y OR N)')STA01990
 1014 FORMAT (/,1X,'MEANS (M) AND VARIANCES (V):',/)           STA02000
 1016 FORMAT (1X,'M',2(1X,I2),' = ',F10.5,5X,'V',2(1X,I2),' = ',F10.6)STA02010
 1020 FORMAT (1X,'PLEASE ENTER VALUES FOR R2 ETC.')            STA02020
 1030 FORMAT (1X,'VALUE FOR R',I3)                             STA02030
 1040 FORMAT (1X,'BEGINNING VALUES FOR R:',/)                  STA02040
 1050 FORMAT (3X,'R',I3,' = ',F8.5)                            STA02050
 1060 FORMAT (/,1X,'CORRECT? (Y OR N)')                        STA02060
 1070 FORMAT (A1)                                               STA02070
 1080 FORMAT (1X,'ERROR...RESPOND Y OR N ONLY')                STA02080
 1085 FORMAT (1X,'ERROR...INVALID R NUMBER')                   STA02090
 1090 FORMAT (1X,'WHICH R DO YOU WANT TO CHANGE?')             STA02100
 1100 FORMAT (I2)                                               STA02110
 1110 FORMAT (1X,'CORRECT R VALUE?')                           STA02120
 1120 FORMAT (1X,'MORE CORRECTIONS? (Y OR N)')                 STA02130
 1130 FORMAT (/,2X,'R',8X,'R',/,1X,'NUM',5X,'VALUE',10X,'B0',12X,'B1',STA02140
     1 12X,'B2')                                                STA02150
 1140 FORMAT (1X,I3,F11.8,F14.4,F14.6,F14.4)                   STA02160
 1150 FORMAT (/,/,                                             STA02170
     1 1X,'ANOTHER ITERATION USING COMPUTED R VALUES? (Y OR N)')STA02180
 1160 FORMAT (1X,'ANOTHER ITERATION USING NEW R VALUES? (Y OR N)')STA02190
 1170 FORMAT (1X,'ERROR...MISSING DATA ON INPUT FILE')         STA02200
      END
```

9.E   PROGRAM 2 — ORIGINAL DATA

FILE: STARPT06 FORTRAN   A   EXXON NUCLEAR - VM/CMS

```
C                                                                       STA00010
C       FORTRAN PROGRAM WRITTEN FOR JOHN JAECH PER HIS INSTRUCTIONS     STA00020
C       RE PUBLICATION :   SPECIAL PROBLEMS IN MEASUREMENT VARIANCE     STA00030
C        ESTIMATION AND HYPOTHESIS TESTING                             STA00040
C     PROGRAM 2 ESTIMATION OF MEASUREMENT PRECISIONS FROM NI BY N ARRAY:STA00050
C       NON-CONSTANT BIAS MODEL (SEC. 8.2)                             STA00060
C                                                                       STA00070
C     INPUT OF NI BY N ARRAY IS FROM A FILE (01)                       STA00080
C       DIMENSIONS OF THE ARRAY IS THE FIRST RECORD IN THAT FILE       STA00090
C       MAX ARRAY SIZE (20,500)                                        STA00100
C                                                                       STA00110
        IMPLICIT DOUBLE PRECISION (A-H,O-Z)                            STA00120
        CHARACTER*1 AY,AN,RESP                                         STA00130
        DIMENSION X(20,500),AVE(20),VAR(20,20),R(20),A(20),B(20),C(20),STA00140
       1 D(20),E(20),F(20),G(20)                                       STA00150
        DATA AY/'Y'/,AN/'N'/                                           STA00160
C                                                                       STA00170
C       READ IN DIMENSIONS (N LT 21,NI LT 51) AND ARRAY X             STA00180
C                                                                       STA00190
        READ (1,*) N,NI                                                STA00200
        IF (N.LE.20.AND.NI.LE.500.AND.N.GT.0.AND.NI.GT.0) GO TO 100    STA00210
        WRITE (6,1000)                                                 STA00220
        STOP                                                           STA00230
  100   DO 110 I=1,NI                                                  STA00240
        IF (N.LE.10) THEN                                              STA00250
           READ (1,*,END=500) (X(J,I),J=1,N)                          STA00260
        ELSE                                                           STA00270
           READ (1,*,END=500) (X(J,I),J=1,10)                         STA00280
           READ (1,*,END=500) (X(J,I),J=11,N)                         STA00290
        END IF                                                         STA00300
        READ (1,1002)                                                 STA00310
  110   CONTINUE                                                       STA00320
C                                                                       STA00330
C       CHECK IF WANT INPUT DATA PRINTED OUT                          STA00340
C                                                                       STA00350
  112   WRITE (6,1004)                                                STA00360
        READ (5,1070) RESP                                            STA00370
        IF (RESP.EQ.AN) GO TO 125                                     STA00380
        IF (RESP.EQ.AY) GO TO 115                                     STA00390
        WRITE (6,1080)                                                STA00400
        GO TO 112                                                     STA00410
C                                                                       STA00420
C       PRINT OUT INPUT DATA                                          STA00430
C                                                                       STA00440
  115   WRITE (6,1008)                                                STA00450
        DO 120 I=1,NI                                                  STA00460
        IF (N.LE.10) THEN                                              STA00470
           WRITE (6,1010) (X(J,I),J=1,N)                              STA00480
        ELSE                                                           STA00490
           WRITE (6,1010) (X(J,I),J=1,10)                             STA00500
           WRITE (6,1010) (X(J,I),J=11,N)                             STA00510
        END IF                                                         STA00520
        WRITE (6,1002)                                                STA00530
  120   CONTINUE                                                       STA00540
C                                                                       STA00550
```

```
FILE: STARPT06 FORTRAN   A    EXXON NUCLEAR - VM/CMS


C       CLEAR MEAN AND VARIANCE ARRAYS                          STA00560
C                                                               STA00570
  125 DO 140 I=1,20                                             STA00580
      DO 130 J=1,20                                             STA00590
      VAR(I,J)=0.0                                              STA00600
  130 CONTINUE                                                  STA00610
      AVE(I)=0.0                                                STA00620
  140 CONTINUE                                                  STA00630
C                                                               STA00640
C       CALCULATE THE MEAN, VARIANCE, AND COVARIANCE OF THE INPUT ARRAY  STA00650
C                                                               STA00660
      SUMAVE=0.0                                                STA00670
      DO 165 I=1,N                                              STA00680
      DO 160 J=1,N                                              STA00690
      IF (J.GE.I) GO TO 145                                     STA00700
      VAR(I,J)=VAR(J,I)                                         STA00710
      GO TO 155                                                 STA00720
  145 SUMI=0.0                                                  STA00730
      SUMJ=0.0                                                  STA00740
      SUMIJ=0.0                                                 STA00750
      DO 150 K=1,NI                                             STA00760
      SUMI=SUMI+X(I,K)                                          STA00770
      SUMJ=SUMJ+X(J,K)                                          STA00780
      SUMIJ=SUMIJ+(X(I,K)*X(J,K))                               STA00790
  150 CONTINUE                                                  STA00800
      SIJSUM=SUMI*SUMJ                                          STA00810
      TEMP=SUMIJ-SIJSUM/NI                                      STA00820
      VAR(I,J)=TEMP/(NI-1)                                      STA00830
      IF (I.EQ.J) AVE(I)=SUMI/NI                                STA00840
  155 CONTINUE                                                  STA00850
  160 CONTINUE                                                  STA00860
      SUMAVE=SUMAVE+AVE(I)                                      STA00870
  165 CONTINUE                                                  STA00880
      AVEBAR=SUMAVE/N                                           STA00890
C                                                               STA00900
C       CHECK IF WANT TO SEE COMPUTED MEAN AND VARIANCE          STA00910
C                                                               STA00920
  166 WRITE (6,1012)                                            STA00930
      READ (5,1070) RESP                                        STA00940
      IF (RESP.EQ.AN) THEN                                      STA00950
        GO TO 168                                               STA00960
      ELSE IF (RESP.NE.AY) THEN                                 STA00970
        WRITE (6,1080)                                          STA00980
        GO TO 166                                               STA00990
      ELSE                                                      STA01000
        CONTINUE                                                STA01010
      END IF                                                    STA01020
C                                                               STA01030
C       PRINT OUT MEAN AND VARIANCE                             STA01040
C                                                               STA01050
      WRITE (6,1014)                                            STA01060
      DO 167 I=1,N                                              STA01070
      WRITE (6,1016) I, AVE(I),I,I,VAR(I,I)                     STA01080
  167 CONTINUE                                                  STA01090
C                                                               STA01100
```

```
FILE: STARPT06 FORTRAN  A    EXXON NUCLEAR - VM/CMS


C      CHECK IF WANT TO SEE CO-VARIANCE                          STA01110
C                                                                STA01120
  168 WRITE (6,1018)                                             STA01130
      READ (5,1070) RESP                                         STA01140
      IF (RESP.EQ.AN) THEN                                       STA01150
        GO TO 174                                                STA01160
      ELSE IF (RESP.NE.AY) THEN                                  STA01170
        WRITE (6,1080)                                           STA01180
        GO TO 168                                                STA01190
      ELSE                                                       STA01200
        CONTINUE                                                 STA01210
      END IF                                                     STA01220
C                                                                STA01230
C      PRINT OUT COVARIANCES                                     STA01240
C                                                                STA01250
      WRITE (6,1020)                                             STA01260
      NLESS1=N-1                                                 STA01270
      DO 170 I=1,NLESS1                                          STA01280
      IPLUS1=I+1                                                 STA01290
      DO 170 J=IPLUS1,N                                          STA01300
      WRITE (6,1022) I,J,VAR(I,J)                                STA01310
  170 CONTINUE                                                   STA01320
C                                                                STA01330
C      DETERMINE WHETHER TO COMPUTE B'S OR SET TO 1.0            STA01340
C                                                                STA01350
  174 WRITE (6,1170)                                             STA01360
      READ (5,1070) RESP                                         STA01370
      IF (RESP.EQ.AY) GO TO 200                                  STA01380
      IF (RESP.EQ.AN) GO TO 175                                  STA01390
      WRITE (6,1080)                                             STA01400
      GO TO 174                                                  STA01410
C                                                                STA01420
C      CALCULATE B'S                                             STA01430
C                                                                STA01440
  175 WRITE (6,1180)                                             STA01450
      DO 195 I=1,N                                               STA01460
      SUM1=0.0                                                   STA01470
      DO 180 J=1,N                                               STA01480
      IF (J.EQ.I) GO TO 177                                      STA01490
      TEMP=LOG(VAR(I,J))                                         STA01500
      SUM1=SUM1+TEMP                                             STA01510
  177 CONTINUE                                                   STA01520
  180 CONTINUE                                                   STA01530
      SUM2=0.0                                                   STA01540
      NLESS1=N-1                                                 STA01550
      DO 190 J=1,NLESS1                                          STA01560
      IF (J.EQ.I) GO TO 187                                      STA01570
      JPLUS1=J+1                                                 STA01580
      DO 185 K=JPLUS1,N                                          STA01590
      IF (K.EQ.I) GO TO 183                                      STA01600
      TEMP=LOG(VAR(J,K))                                         STA01610
      SUM2=SUM2+TEMP                                             STA01620
  183 CONTINUE                                                   STA01630
  185 CONTINUE                                                   STA01640
  187 CONTINUE                                                   STA01650
```

```
FILE: STARPT06 FORTRAN   A     EXXON NUCLEAR - VM/CMS


    190 CONTINUE                                                       STA01660
        SLB=(SUM1/N)-(2*(SUM2/(N*(N-2))))                             STA01670
        B(I)=EXP(SLB)                                                 STA01680
        A(I)=AVE(I)-(B(I)*AVEBAR)                                     STA01690
        WRITE (6,1200) I,B(I),I,A(I)                                  STA01700
    195 CONTINUE                                                      STA01710
        GO TO 210                                                     STA01720
C                                                                     STA01730
C       SET B'S TO 1.0                                                STA01740
C                                                                     STA01750
    200 WRITE (6,1190)                                                STA01760
        DO 205 I=1,N                                                  STA01770
        B(I)=1.0                                                      STA01780
        A(I)=AVE(I)-(B(I)*AVEBAR)                                     STA01790
        WRITE (6,1200) I,B(I),I,A(I)                                  STA01800
    205 CONTINUE                                                      STA01810
        GO TO 210                                                     STA01820
C                                                                     STA01830
C       GET INITIAL VALUES FOR R1 THROUGH RN                         STA01840
C                                                                     STA01850
    210 WRITE (6,1002)                                                STA01860
        DO 220 I=1,N                                                  STA01870
        WRITE (6,1030) I                                              STA01880
        READ (5,*) R(I)                                               STA01890
    220 CONTINUE                                                      STA01900
C                                                                     STA01910
C       ECHO BACK INPUTTED R VALUES FOR USER VERIFICATION            STA01920
C                                                                     STA01930
    230 WRITE (6,1040)                                                STA01940
        DO 240 I=1,N                                                  STA01950
        WRITE (6,1050) I,R(I)                                         STA01960
    240 CONTINUE                                                      STA01970
    250 WRITE (6,1060)                                                STA01980
        READ (5,1070) RESP                                           STA01990
        IF (RESP.EQ.AY) GO TO 310                                     STA02000
        IF (RESP.EQ.AN) GO TO 270                                     STA02010
        WRITE (6,1080)                                                STA02020
        GO TO 250                                                     STA02030
C                                                                     STA02040
C       ALLOW USER TO CORRECT INPUTTED R VALUES                      STA02050
C                                                                     STA02060
    270 WRITE (6,1090)                                                STA02070
        READ (5,*) I                                                  STA02080
        IF (I.LE.N.AND.I.GE.1) GO TO 280                              STA02090
        WRITE (6,1075)                                                STA02100
        GO TO 270                                                     STA02110
    280 WRITE (6,1110)                                                STA02120
        READ (5,*) R(I)                                               STA02130
    290 WRITE (6,1120)                                                STA02140
        READ (5,1070) RESP                                           STA02150
        IF (RESP.EQ.AY) GO TO 270                                     STA02160
        IF (RESP.EQ.AN) GO TO 230                                     STA02170
        WRITE (6,1080)                                                STA02180
        GO TO 290                                                     STA02190
C                                                                     STA02200
```

FILE: STARPT06 FORTRAN A   EXXON NUCLEAR - VM/CMS

```
C     CALCULATE R0                                          STA02210
C                                                           STA02220
  310 D2=0.0                                                STA02230
      D3=0.0                                                STA02240
      S1D4=0.0                                              STA02250
      S2D4=0.0                                              STA02260
      DO 350 I=1,N                                          STA02270
      D2=D2+((B(I)**2)/R(I))                                STA02280
      D3=D3+(VAR(I,I)/R(I))                                 STA02290
      SQB=B(I)**2                                           STA02300
      DO 340 J=1,N                                          STA02310
      IF (J.EQ.I) GO TO 320                                 STA02320
      S1D4=S1D4+((SQB*VAR(I,J))/(R(I)*R(J)))                STA02330
  320 IF (I.GT.NLESS1) GO TO 330                            STA02340
      IF (J.LE.I) GO TO 330                                 STA02350
      S2D4=S2D4+((B(I)*B(J)*VAR(I,J))/(R(I)*R(J)))          STA02360
  330 CONTINUE                                              STA02370
  340 CONTINUE                                              STA02380
  350 CONTINUE                                              STA02390
      D4=S1D4-(2.0*S2D4)                                    STA02400
      R0=((NI-1)*((D2*D3)-D4))/(NI*D2**2)                   STA02410
      WRITE (6,1210) R0                                     STA02420
C                                                           STA02430
C     CALCULATE NEW R VALUE                                 STA02440
C                                                           STA02450
      WRITE (6,1002)                                        STA02460
      DO 355 I=1,N                                          STA02470
      C(I)=B(I)*SQRT(R0)                                    STA02480
      D(I)=C(I)**2                                          STA02490
      WRITE (6,1220) I,C(I),I,D(I)                          STA02500
  355 CONTINUE                                              STA02510
      WRITE (6,1002)                                        STA02520
      DO 405 I=1,N                                          STA02530
      FSUM1=0.0                                             STA02540
      FSUM2=0.0                                             STA02550
      GSUM1=0.0                                             STA02560
      GSUM2=0.0                                             STA02570
      GSUM3=0.0                                             STA02580
      E(I)=1.0                                              STA02590
      DO 400 J=1,N                                          STA02600
      IF (J.EQ.I) GO TO 380                                 STA02610
      E(I)=E(I)+((C(J)**2)/R(J))                            STA02620
      FSUM1=FSUM1+(VAR(J,J)/R(J))                           STA02630
      FSUM2=FSUM2+((C(J)*VAR(I,J))/R(J))                    STA02640
      GSUM1=GSUM1+(VAR(J,J)/R(J))                           STA02650
      GSUM2=GSUM2+(((C(J)**2)*VAR(J,J))/(R(J)**2))          STA02660
      IF (J.EQ.N) GO TO 380                                 STA02670
      JPLUS1=J+1                                            STA02680
      DO 360 K=JPLUS1,N                                     STA02690
      IF (K.NE.I) GSUM3=GSUM3+((C(J)*C(K)*VAR(J,K))/(R(J)*R(K)))  STA02700
  360 CONTINUE                                              STA02710
  380 CONTINUE                                              STA02720
  400 CONTINUE                                              STA02730
      F(I)=(E(I)*VAR(I,I))+(D(I)*FSUM1)-(2.0*C(I)*FSUM2)    STA02740
      G(I)=(E(I)*GSUM1)-GSUM2-(2.0*GSUM3)                   STA02750
```

```
FILE: STARPT06 FORTRAN  A    EXXON NUCLEAR - VM/CMS

      R(I)=(((NI-1)*((E(I)*F(I))-(D(I)*G(I))))/(NI*(E(I)**2)))    STA02760
     1 -(D(I)/E(I))                                              STA02770
      DO 402 J=1,N                                               STA02780
      IF (I.EQ.J) GO TO 401                                      STA02790
  401 CONTINUE                                                   STA02800
  402 CONTINUE                                                   STA02810
      WRITE (6,1230) I,R(I),I,E(I),I,F(I),I,G(I)                 STA02820
  405 CONTINUE                                                   STA02830
C     GO TO 440                                                  STA02840
C                                                                STA02850
C     CHECK IF WANT MORE ITERATIONS                             STA02860
C                                                                STA02870
  410 WRITE (6,1150)                                             STA02880
      READ (5,1070) RESP                                         STA02890
      IF (RESP.EQ.AY) GO TO 310                                  STA02900
      IF (RESP.EQ.AN) GO TO 420                                  STA02910
      WRITE (6,1080)                                             STA02920
      GO TO 410                                                  STA02930
  420 WRITE (6,1160)                                             STA02940
      READ (5,1070) RESP                                         STA02950
      IF (RESP.EQ.AY) GO TO 210                                  STA02960
      IF (RESP.EQ.AN) GO TO 440                                  STA02970
      WRITE (6,1080)                                             STA02980
      GO TO 420                                                  STA02990
  440 STOP                                                       STA03000
C                                                                STA03010
C     ERRORS                                                     STA03020
C                                                                STA03030
  500 WRITE (6,1240)                                             STA03040
      STOP                                                       STA03050
C                                                                STA03060
C     FORMATS                                                    STA03070
C                                                                STA03080
 1000 FORMAT (1X,'ERROR ON ARRAY DIMENSIONS')                   STA03090
 1002 FORMAT (1X)                                                STA03100
 1004 FORMAT (/,1X,'DO YOU WANT TO SEE THE INPUT DATA? (Y OR N)')STA03110
 1008 FORMAT (/,1X,'INPUT DATA:',/)                             STA03120
 1010 FORMAT (1X,10F7.3)                                         STA03130
 1012 FORMAT (/,1X,'DO YOU WANT TO SEE THE MEAN AND VARIANCE? (Y OR N)')STA03140
 1014 FORMAT (/,1X,'MEAN (M) AND VARIANCE (V):',/)              STA03150
 1016 FORMAT (1X,'M',1X,I2,' = ',F10.5,5X,'V',2(1X,I2),' = ',F10.6)STA03160
 1018 FORMAT (/,1X,'DO YOU WANT TO SEE THE COVARIANCE? (Y OR N)')STA03170
 1020 FORMAT (/,1X,'COVARIANCES (C) .',/)                       STA03180
 1022 FORMAT (1X,'C',2(1X,I2),' = ',F10.6)                      STA03190
 1026 FORMAT (/,1X,'PLEASE ENTER VALUES FOR R1 ETC.',/)         STA03200
 1030 FORMAT (1X,'VALUE FOR R',I3)                              STA03210
 1040 FORMAT (/,1X,'BEGINNING VALUES FOR R:',/)                 STA03220
 1050 FORMAT (3X,'R',I3,' = ',F8.5)                             STA03230
 1060 FORMAT (/,1X,'CORRECT? (Y OR N)')                         STA03240
 1070 FORMAT (A1)                                                STA03250
 1075 FORMAT (1X,'ERROR...INCORRECT R NUMBER...TRY AGAIN')      STA03260
 1080 FORMAT (1X,'ERROR... RESPOND Y OR N ONLY')                STA03270
 1090 FORMAT (1X,'WHAT R DO YOU WANT TO CHANGE?')               STA03280
 1100 FORMAT (I2)                                                STA03290
 1110 FORMAT (1X,'CORRECT R VALUE?')                            STA03300
```

FILE: STARPT06 FORTRAN   A   EXXON NUCLEAR - VM/CMS

```
1120 FORMAT (1X,'MORE CORRECTIONS? (Y OR N)')                          STA03310
1130 FORMAT (/,2X,'R',8X,'R',/,1X,'NUM',5X,'VALUE',7X,'B0',9X,'B1',    STA03320
    1 9X,'B2')                                                         STA03330
1140 FORMAT (1X,I3,3X,F8.6,3X,F8.1,3X,F8.3,3X,F8.1)                    STA03340
1150 FORMAT (/,/,                                                      STA03350
    1 1X,'ANOTHER ITERATION USING COMPUTED R VALUES? (Y OR N)')        STA03360
1160 FORMAT (1X,'ANOTHER ITERATION USING NEW R VALUES? (Y OR N)')      STA03370
1170 FORMAT (/,1X,'DO YOU WANT ALL BS SET TO 1.0 ? (Y OR N)')          STA03380
1180 FORMAT (/,1X,'BS WILL BE CALCULATED',/)                          STA03390
1190 FORMAT (/,1X,'BS ARE SET TO 1.0',/)                              STA03400
1200 FORMAT (1X,'B',I3,' = ',F8.4,5X,'A',I3,' = ',F8.4)               STA03410
1210 FORMAT (/,/,1X,'R0 = ',F10.6)                                    STA03420
1220 FORMAT (2X,'C',I3,' = ',F10.6,5X,'D',I3,' = ',F10.6)             STA03430
1230 FORMAT (1X,'R',I2,' = ',F14.8,/,1X,'E',I2,' = ',F14.4,/,         STA03440
    1 1X,'F',I2,' = ',F14.6,/,1X,'G',I2,' = ',F14.4,/)               STA03450
1240 FORMAT (1X,'ERROR ON INPUT DATA FILE...PLEASE CORRECT AND RERUN.')STA03460
    END                                                               STA03470
```

INDEX